Out of Time

OUT OF TIME

The Intergenerational Abduction Program Explored
(Second Edition)

STEVE ASPIN

Grosvenor House
Publishing Limited

All rights reserved
Copyright © Steve Aspin, 2023
First edition 2022

The right of Steve Aspin to be identified as the author of this work has been asserted in accordance with Section 78 of the Copyright, Designs and Patents Act 1988

The book cover is copyright to Steve Aspin
Cover Illustration: *Fractured Disc* by Jared S. Tarbell, copyright 2022

This book is published by
Grosvenor House Publishing Ltd
Link House
140 The Broadway, Tolworth, Surrey, KT6 7HT.
www.grosvenorhousepublishing.co.uk

This book is sold subject to the conditions that it shall not, by way of trade or otherwise, be lent, resold, hired out or otherwise circulated without the author's or publisher's prior consent in any form of binding or cover other than that in which it is published and without a similar condition including this condition being imposed on the subsequent purchaser.

A CIP record for this book
is available from the British Library

Paperback ISBN 978-1-80381-521-3
Hardback ISBN 978-1-80381-522-0
eBook ISBN 978-1-80381-523-7

Out of Time Books
www.outoftimebook.info

Many abductees have shared their stories with me over the past 15 years. Most have made the choice to remain silent about their ordeal, keeping it from husband, wife, partner, family and friends, fearing they would be disbelieved and that their relationships, careers or reputation would suffer as a consequence.

The experiences and insights of these courageous people, who each discover their lives to be entangled with this phenomenon, are woven discretely into the narrative fabric of *Out of Time*. They remain anonymous but know who they are. In recognition of their steadfast determination to hold their lives together in the face of the 'secret life' they all share, and to the millions of aware or partially aware abductees around the world who feel they have no one to confide in, this book is dedicated to them all.

Included in this dedication are my mother, maternal grandmother and great-grandmother, whose lives were caught up with this intrusive program and who are now gone from this world.

Contents

Foreword ix
by Don Donderi, PhD

Introduction 1

Prologue 7
One July Morning ...

Chapter One: The Slow Revelation 13
A short factual history of our involuntary bonds with non-human entities

Chapter Two: A Spectrum of Early Cases 45
Which came to widespread public attention and collectively confirmed what is going on

Chapter Three: The Program 69
What is it, and why describe it as a 'Program'?

Chapter Four: Physical Evidence 91
What they leave behind: scars and scoops, implants and hair

Chapter Five: They're Doing What? 115
Puzzlin' evidence: get used to it ...

Chapter Six: The Great Pioneer 137
Personal time spent with Budd Hopkins during his final years

Chapter Seven: The Assiduous Professor 161
My invaluable work with Dr David Jacobs and our growing personal friendship

Chapter Eight: Question Time 181
Is the Program really real? How long has it been going on? – and other questions

Chapter Nine: "They" 201
Where do they come from? Why are different beings described? What's the endgame?

Epilogue 227

Appendix A 233

Appendix B 235

Appendix C 239

Appendix D 255

Bibliography 297

Notes and References 307

Acknowledgements 319

About the Author 321

Foreword

by Don Donderi, PhD

Out of Time—Time for What? Humans have had plenty of time—some 300,000 years—to evolve and settle on Earth, where we have become the dominant species. Here, for better or worse, we have run things our own way, surviving both natural catastrophes and our own self-destructive behavior. But the cumulative evidence suggests that we may soon be elbowed aside by extraterrestrials (ETs) who plan to dominate Earth, managing both the planet and us in their own interest.

That is the essential conclusion of Steve Aspin's *Out of Time*. Why does Steve believe this is even possible? Because he has experienced covert interference with his own life since infancy and, in later years, has sought out many dedicated researchers in an effort to understand what has been learned about the UFO/UAP and abduction phenomena. His own experiences, and his review of what the rest of us have learned, are reported clearly and succinctly in this book.

Steve was born into a close-knit family in the North of England. 'Anomalous experiences' were an irregular but routine part of his life and he discovered, as he grew to maturity, that similar experiences seemed to be common to one direct ancestral line of his family. The possible objective of this widely reported 'intergenerational' aspect of the phenomenon is one of the core issues explored in *Out of Time*. These intrusions have long been accepted as part of folklore, of the supernatural, of mischievous 'pixies'; but from an early age, Steve became aware that this phenomenon—and, indeed, some traditional folklore—may originate from very real circumstances to which we all should be paying much closer attention.

Having received a good education, Steve went on to found and manage an international company that kept him on the road and in the air over much of the world for many years. As an international traveler, he made himself a part of the world-wide community of people who study and report on the UFO and abduction phenomena. *Out of Time* summarizes his own encounters and reviews those of others who, driven by personal necessity or curiosity, have documented their own experiences, or researched and described those of others. *Out of Time* tells us that it may be past time to recognize and respond to a potential threat in our near future that we must not ignore.

I am a retired McGill University professor with a PhD in experimental psychology from Cornell University. My research has been focused on understanding human visual perception and memory—gateways through which human experience is obtained and recorded. I practice as a consultant in the fields of ergonomic design, transportation, and information technology. I have written or co-authored two psychology textbooks and more than one hundred peer-reviewed research papers and technical reports. I own a design patent and a design copyright. My professional knowledge reinforces the conclusion that Steve Aspin may be right.

My interest in the UFO/UAP evidence began in 1965 when witnesses, not for the first time, reported UFO sightings across North America. Witness reports are based on visual perception and memory—my own research interests—and are often supplemented by instrumental evidence. I wrote two books about this evidence, both supported by my professional understanding of vision and memory: *UFOs, ETs and Alien Abductions: A Scientist Looks at the Evidence* (Hampton Roads Press, 2013) and *Truth, Lies and ETs: How we Stumbled into the Universe* (Moonshine Cove Press, 2022).

Out of Time describes Steve Aspin's direct experience of this widely reported phenomenon. It reviews the work of writers of our generation whom he knew personally, and the work of people who began to tell us about UFOs and 'alien abductions' many years ago. *Out of Time* is clear, thorough, and accurate. Read it, because we all need to learn why we may soon be *Out of Time*.

Don C. Donderi, PhD,
Associate Professor (Retired), McGill University, Montreal

Introduction

The modern phenomenon of unidentified flying objects, as reported over decades by thousands of military and civilian pilots and witnessed by citizens of every kind all over the Earth, has been with us now for over a century. The issue has been successfully normalised in popular culture to become background noise to the point where most people consider it to be of little importance, relegating the subject to a minor comic footnote in their busy lives. In many cases, the question of what these extraordinary things seen in the sky might be, or their significance for humanity, is either disregarded or completely ignored as irrelevant to people's everyday concerns.

This subject is effectively marginalised in societal consciousness for one reason above all others: the evident *lack of intentionality* of those controlling these objects, which are generally assumed to be extraterrestrial. They have been seen by millions of people; they have been recorded on military and civilian radar around the globe; they are obviously under intelligent control and easily outperform all our military aircraft; they occasionally land and leave measurable physical traces on the ground.

But they don't "land on the White House lawn."[1] They make no attempt to communicate with our political and societal leaders in any way—at least in any way which is open and public—although they might easily choose to do. Why not? What have they been doing here after all these decades? There is an apocryphal story from the 1940s when, after President Harry Truman had been briefed about the UFO issue and told that the craft were believed to be of extraterrestrial origin, he asked his intelligence chiefs, "So what do the sons of bitches want?" This book will attempt to offer some answers to this question.

Publicly recognised by the international press media since the Spring of 1947,[2] this issue is only now being officially acknowledged as worthy of attention with some honesty and openness in the United States after decades of obfuscation and denial that it even exists or is of any consequence.³ The lifetime experiences of thousands of people who have suffered close-up interaction with this phenomenon are finally being acknowledged, albeit slowly and reluctantly, because evidently this thing ain't goin' away. Many wish it would, but it looks like 'they'—whoever and whatever 'they' might be—are here to stay.

The testimonies of those who report interactions with the occupants of these craft span the spectrum of highly credible and evidence-driven, through the deeply disturbing, to the kooky and blatantly delusional. 'Confessional abductee literature' is a recognised genre within the landscape of the extraordinary world which seeks to explore and understand this UFO/UAP phenomenon. The book you are about to read is not intended to be of that genre because it is not exclusively an account of one or more witness' personal experiences and memories of encounters. There have been many such published works over the past fifty years and the fundamentals of the phenomenon are now well understood and recognised. Rather, the personal angle is deployed in this book as a doorway leading into a deeper enquiry as to what this phenomenon might really be about: what is going on here, and is it more important to us than we have been conditioned to think by long familiarity with the apparent non-action from these visitors?

Chapter One is a broad-ranging chronological narrative serving as just such an entry point to the exploration of more specific questions which follow. The experiences of a child born in England in 1956 who, throughout his formative years, experienced repeated encounters which he did not understand and for which his family, school environment and the society at large offered no explanation, offer a window into a world of strangeness and forge a pathway through it all. The reader should bear in mind that this account is written retrospectively from a position of greater understanding gained in later life. At the time, almost none of it made any sense. But now, though answers to many of the deeper questions yet remain elusive, some clarity obtains.

It is important to emphasise that, although I later worked extensively with the highly assiduous Dr David Jacobs, who helped me to successfully recover deeper memories buried from some of the encounters outlined in Chapter One, the narratives related here are exclusively those retained in normal memory at the time, so lacking many details buried in 'missing time'. This decision was made for two reasons:

- To prevent, or at least deter, any criticism about the effective use of hypnosis to assist with memory recovery,[4] as none of the encounters uncovered or further clarified by hypnosis are used in the narrative

- To offer the reader an accurate-as-possible guide to precisely what was experienced, and remembered, *at that time*, and which led to the mid-life decision to seek out assistance in understanding the implications of these memories at a deeper level

It is often admitted that when one chooses to write a book on this subject, there is invariably a second book's worth of material which has by choice or necessity been excluded. So it is with this one: a second volume detailing all the remembered experiences could be written. This might contain a great deal more detail of interactions with the abductors, but little—although not nothing—which the reader already interested in the subject has not previously encountered elsewhere. The possible exception might be the extensive interactions with alien-human hybrids, now the dominant phase of the program and which the current volume touches on only peripherally. It's a big subject but needs a good grounding in the basics which this book, among other things, attempts to offer.

From Chapter Two through to the conclusion of the book, we shall take a journey together through this phenomenon and try to make sense of it, piece by piece. Readers are cautioned that their assumptions may on occasion be challenged. Many of the author's personal ideas and discoveries are likely to prove controversial. However, my direct personal experience is rarely the exclusive focus of the narrative: rather, specific examples are deployed to shed light on more generally pervasive aspects of the matter under scrutiny.

The phenomenon is complicated and truly 'alien' to the thought processes embedded in our intellectual culture and inherited belief systems, but it nevertheless has its own internal logic, and it is possible to come to understand in broad terms what 'they' are doing here. The timescale of the intergenerational program may be long, but it is finite and goal-directed[5] and, I believe, its fundamental architecture may be understood. The 'Why?' question remains unanswered and the subject of speculation, of which this field of enquiry has no deficit.

As a light relief from the serious subject matter, the reader will I hope forgive occasional brief excursions into observations stimulated by my travel experiences, not just in the USA—I'm not American and am continually surprised by what I discover during my extensive travels there over the years—but also to other destinations, many of which would never have been visited were it not for the insistent intrusion of this phenomenon throughout life impelling me towards new frontiers of experience in an attempt to understand it and to 'cover all the bases'.

I hope at least that the reader's interest in exploring the subject may be aroused to the degree that s/he undertakes further research into this little understood but vitally important matter.

A Personal Note

The incomparable Budd Hopkins kept a personal journal during the last two years of his life, before he passed away in August 2011. Although much of its content is intended for his immediate family and close friends only, some passages are of general interest and may be shared more broadly.

There are observations on late life and artistic endeavour. This passage was written on 23 January 2011:

> As I've mentioned before, I've often thought about the issues raised by my writing this journal, a posthumous effort … a very late work … but for what purpose? Despite my casual attitude to its composition, I can't fail to regard it as a kind of literary effort. Having published four books and written many articles, it's almost reflexive of me to think of this journal as a kind of bastard artwork … I was intrigued by a piece in today's *Times Book Review* section about artists' late work (by the way, the *Times Book Review* seems to be moving away from its committed middle-brow status and becoming, itself, more seriously literary).
>
> The review I read was *Lastingness—the Art of Old Age*, by the novelist Nicholas Delbanco. It was written by Brooke Allen, about whom I know nothing though his review is insightful. About the phenomenon of a late style, a more terse, stripped down version of an artist's earlier work, he has this to say: "In youth," he quotes Delbanco as saying, "it's the reception of the piece and not its production that counts. But to the ageing writer, painter or musician, the production can signify more than result; it no longer seems important that the work be sold."
>
> It is a profound observation: with time and age, the act of showing becomes increasingly subordinate to the act of making, and gratification turns ever further inward … but this is surely not the only reason for the concentrated effect of the late style. The simple specter of mortality must count for something …

These incisive, poignant observations about late-in-life creativity and artistic effort sum up quite succinctly my attitude to both writing this book, and its reception. The act of *showing* is without question subordinate to the act of *making*: it is more important to me that I write all this down for posterity than it is of concern that the finished work gains a large readership. But this is not the only motivation. I am sixty-six years old at the time of writing, and the 'spectre of mortality' definitely counts for something. You might say that I'm running *Out of Time*.

If even only a few curious people read this work, enjoy the experience, and learn something about this intrusive phenomenon affecting millions of people all over Planet Earth, the effort will have been worthwhile. If you suspect that you may be among the one-in-twenty or one-in-fifty to whom the narrative has special relevance (see Chapter Eight), I hope this book may offer some reassurance and help guide you on your way.

Prologue

One July Morning ...

When I was sixteen years old, I took a temporary job working at an agricultural nursery in Cheshire, in northwest England. In its rich and fecund soil, the owners had built a successful nationwide business planting, growing, and selling rose trees by the thousand in every variety and coloration then known to the rose-growing world.[1]

From late June to late August 1972 I worked at the rose nursery from Sunday through Friday, cycling the seven-mile journey every morning to arrive prior to the 8am clock-in time and leaving after clocking-out at 5:30 to cycle home. Almost all the temporary summer staff were school/college students between sixteen and twenty years old like me, working the summer holiday under open skies to earn some extra money. As the school leaving age was by then 16 years, few of my fellow-workers were younger than me. As I was a non-smoker, only very occasional drinker (the legal age to buy alcohol was eighteen but I easily passed muster) and cycled everywhere on my French road-racing bike, I was super-fit and full of youthful energy, so I easily held my own in this physically demanding job.

We worked to the soundtrack of the Top 20 singles played on portable transistor radios (remember them?). Regaling us from field to field were *School's Out* by Alice Cooper, *Sylvia's Mother* by Dr Hook and The Medicine Show, and *Starman* by David Bowie. This was the golden era of 45rpm vinyl singles. The iPod, iPad, and iPhone were decades in the future; there were no cell phones, no internet and no social media: happier and freer times.

As I was generally out Friday night with friends after working for six straight days in the fields, on Saturday mornings I slept late. Saturday was the only day when I didn't get up at 6:45 to cycle to the nursery to work in the fields of rose trees and was therefore a precious time. Saturday mornings were special.

On those afternoons I hung out with friends in the cafes and music shops in the nearby historic city of Chester. On Sunday I returned to the nursery, where double-money (!) was offered for a slightly shorter 8am to 4pm working day.

So far, so ordinary. Commonplace; unremarkable. There is a reason why this regular 7-day routine is important, which will soon be revealed.

During the late evening of Thursday, 20th July 1972 I was overcome by a strong impulse to wake up early the next Saturday morning because I was to 'go somewhere'. This perplexing compulsion was uncommon, to say the least. But, for some reason it seemed natural so, rather than question it I actually looked forward to finding out to where I was supposed to go.

By 5am (British Summer Time = UT+1 hour) in July at this latitude, the Sun is risen and the morning bright. Right on cue, on Saturday morning 22nd July 1972 I awoke with a start with no alarm call at 5am, got dressed and went out the front door into the bright July morning. As it was so early on a Saturday morning, no one was around in the middle-class suburban neighbourhood where we lived. The morning temperature shortly after sunrise was around 20C under a cloudless blue summer sky.

About a half-mile from home, some new houses were being built. The unfinished structures had exterior walls and most had roof timbers but no windows, doors, nor inner fittings. Building site materials, like piles of bricks and cement lay around. This site was set back from a local country road, the B5463, and perhaps 200m from the nearest inhabited dwelling. One can still see these long-completed houses today beside the B5463, now inhabited by a mixed community of settled middle-class folks.

I walked quickly to this quiet, still place on this bright July morning arriving around 5:45. I remember thinking, "OK, I'm here. What now?"

Suddenly I found myself in a different area of the part-built houses, looking at my wristwatch, the strap of which had come loose. (Surely, it was secure when I'd left home.) To my surprise it displayed 8:05. The day was noticeably warmer and brighter, the Sun higher in the sky, and the distant buzz of traffic now audible. I was dimly aware that, despite the seamless 'missing time', a great deal seemed to have just happened. It was as if I had suffered some sort of mental blackout.

Colours in the blue/indigo part of the spectrum filled the air, accompanied by buzzing and whooshing noises in my ears. I distinctly remember a voice telling me (inside my head) that I was going to forget everything about the 'journey' this morning. For decades I tried to recall what had happened during the missing two hours and twenty minutes but could remember only the end of the experience with the blue-spectral lights and the voice, and always felt uneasy knowing the seamless missing time had vanished from memory.

A striking spectacle visible through the empty window-opening in the naked new red brick in front of me was a 'flying saucer'. It was clear, bright and in full view about 100m away at approximately 30 degrees of elevation from the horizontal line-of-sight. This memory is vivid because it was so unexpected, so unlikely. I had never seen a UFO before but recognised it immediately for what it apparently was from photos and drawings in popular culture and films. Indeed, I had bought an illustrated magazine titled, *The Flying Saucer Menace*,[2] at a newsagent some five years earlier. It hovered about 30m off the ground and looked like it was somehow in a different zone or dimension to everything else. This is difficult to explain, but if had you seen it you'd understand: it looked as though it was somehow *out of time*. The hovering disc[3] was directly in my line of vision, slowly drifting from left to right. With the disorientation, sudden dislocation of time, insistent buzzing and whooshing sounds and the blue-spectral lights all around, I had absolutely no idea what was going on. Believe it or not, at first it seemed to me that the hovering disc might have nothing whatsoever to do with my current predicament but was there just by coincidence, i.e. a random UFO sighting. Now, I know better.

Few people in northwest England had heard of alien abductions in 1972. It's possible that perhaps one in a hundred people might have read about the 1957 Antônio Villas-Boas case in Brazil, or Betty and Barney Hill's September 1961 encounter when driving through the White Mountains in New Hampshire. Both these cases had gained some notoriety following the publication of John Fuller's book, *The Interrupted Journey*, in January 1966. But these oddball cases had been all but absent from the mainstream news media, confined to a small subset of the population interested in such strange things. Even among UFO enthusiasts, I later discovered, they were regarded with scepticism and incredulity. The 'abduction phenomenon' was, and is still, among our prevailing scientific and socio-cultural paradigms, considered to be unlikely and improbable beyond question.

But it certainly looked as though it might indeed be real—and was happening to me!

The 'voice' was distinctive, with a sound quite deep and human-like. It was not unfriendly, just matter-of-fact, almost apathetic. Yet there was no actual sound in the air; the voice simply filled up my head and the understanding was,

"It doesn't matter what we tell you about all this, you're not going to remember anything anyway." I do remember responding something like, "Oh, yes—I will remember! I'll never forget this!" but indeed they were correct about the memory: the full details were deliberately and effectively blocked, most likely buried so deep in the long-term memory as to become inaccessible.[4]

Right after the blue-spectral lights, the buzzing/whooshing sounds, and the appearance of the hovering disc, everything suddenly came 'back to normal' just like snapping your fingers. I found myself thinking, "Right, now that's over, time to go home," followed by, "What the heck happened there? How could I be missing two hours and twenty minutes just like that? Where did that time go? What was I doing? And why did I get out of bed this early on a Saturday morning and come here anyway?" A persistent, deeply uneasy feeling came over me.

There was a hot burning sensation on the cheekbone below my left eye socket. It felt damp, like it was weeping. I looked in the bathroom mirror when I got home and saw a perfectly round, red mark of raw skin, as though it had been burned or scraped off with a fine scalpel. The mark was the diameter of a large coin. Within a day or so it scabbed over: the scab became hard, brown and scaly, then faded to almost the colour of the surrounding skin but slightly darker and continued to feel rough to the touch. After about three months or so it eventually flaked off and disappeared. At sixteen years old, you tend to be very concerned with your appearance, and any abnormal mark like this on your face can be a big issue, but surprisingly I wasn't too bothered about it and rarely gave it much thought. Only in 2009 did I meet another abductee—a senior policeman living in northern England, and around twenty years younger than me—who had suffered an identical coin-sized facial burn following an abduction event. He reported that the mark on his face had been visible for three months, as with mine, and had gone through roughly the same appearance cycle prior to its eventual disappearance.

I didn't talk about this incident with anyone for years. Partly this was because I remembered only small bits of it and had no memory of what occurred during the missing two hours and twenty minutes, and also because I didn't understand what on Earth could have happened. I wouldn't know what to say. People might think I needed psychiatric help—rather taboo in the provincial culture at the time. I knew that I was not crazy but wasn't going to risk the consequences that revealing it might very well generate. In 1972, in the English provinces, what happened to me wasn't supposed to be real, so the only explanation I might look forward to was likely to be 'psychiatric'.

Surprising as it may seem, I pushed the whole incident to the back of my mind and locked it up for years because, what do you do with it? How to make any sense of it? I got on with teenage life, rock music, college work and planning to leave home for university in a different region of the country, and all the rest.

But memory of this incident lurked in the unconscious for thirty-five years. The weird 'missing time'; the close-up UFO sighting; the circular burn on the cheek—all obviously somehow connected. Not to mention my compulsive determination to get up at 5am on that morning and walk to a deserted location ten minutes from home. When the event pushed back into memory from time to time as an adult, I eventually did talk about it to a few people.

This was the crest of one high wave of memory in a dark ocean of mostly hidden experiences of this phenomenon.

In taking me on a Saturday morning in late July of 1972, in the specific way they did, they revealed that they know exactly what they—and we—are doing. Not only did they know to abduct me precisely on the one day of the week and at the one time of day when I was least likely to be missed, but they somehow placed in my mind the determination to rise from sleep at that specific early hour and walk for 10 minutes to the collection point. Why not take me from home, as I now know now they had done on numerous occasions throughout my childhood?

There are reasons why they sometimes do things exactly as they did on that day, because there is almost nothing random about this phenomenon. It looks like a plan may be at work.

Chapter One

The Slow Revelation

*A short factual history of our involuntary bonds
with non-human entities*

First Things First ...

I was born in April 1956 and had, superficially, an ordinary childhood with a functional nuclear family-of-origin in the English provinces, moving geographically several times between the ages of two and nine and attending normal schools.

My father was a career academic. With a university degree in French, he was a fluent speaker and literate written communicator in that language and following discharge from Army service a year after I was born, he became a high school teacher. His hobby and passion for as long as I can remember was buying and collecting old French books, the earlier and rarer the better and especially first editions if they had beautiful and exotic leather bindings. Later his collecting mania expanded to embrace a wider spectrum of antiquarian books, including classic English literature and early bibles in several languages. By his early 40s he was Professor of French at Liverpool University, with a few intermediate career moves on the way. As he changed jobs between one academic year (September to early July) and the next we relocated to a new part of the country every second or third year, always during the long summer holidays, to accommodate Dad's career ambitions. Between 1956 and 1965 we had moved from West Yorkshire where I was born, across The Pennines to Lancashire, to northeast Hertfordshire near Cambridge, to Cleveland in the northeast close to the sea where lie my most cherished memories of a near-ideal childhood.

In the summer of 1965, when I was nine years old and my younger sister was seven, we moved from Guisborough in Cleveland to Cheshire in northwest England where my father settled into a lecturing job (referred to as *tenure* in the USA) at a teacher training college in Chester for several years, prior to his final career move to the nearby Liverpool University twenty miles away. All this time he also worked the home-based business he'd built up over the years trading in rare, high-value antiquarian books. From 1965 to 1974 we lived in the same area on the Wirral Peninsula, only moving once locally to upgrade to a larger and fancier house, so my formative teenage years were all spent on the Wirral. Until moving south in 1974, I spoke with a similar accent to one of The Beatles, who were all local Liverpool lads.

When we were young, my mother was predominantly a full-time homemaker looking after us but she later supplemented the household income by teaching office secretarial skills to teenage girls in their final year of school, so in these years both parents were education professionals.

However, something else was always going on with my mother and me (and my grandmother and great-grandmother too, but they didn't live with us) which was far from 'ordinary'. It is this lifelong phenomenon and its sometimes highly intrusive effects on our lives which finally motivated me, in my mid-60s, to write this book before I move on to the Great Beyond.

Night Intrusions

As a child there were many bizarre night-time incidents, none of which were accommodated inside what was offered by society at large as 'shared consensus reality'. There was sometimes blue light in the house at night. I remember being surrounded by (usually three) small, spindly creatures and unable to move to escape them. There was a distinct smell and feel to them, not pleasant to be around. The paralysis and inability to move voluntarily on these occasions often seemed to last for long hours, yet I knew I was fully awake. I remember being moved around and then 'examined' while lying on my back on a hard surface; sometimes needles and other intrusive instruments were deployed.

I have no certainty of the frequency of these night-time visitations but they happened, it seems now, every few weeks throughout the years of my childhood, certainly from age seven, and probably prior to that. Researchers have discovered that abductions begin in infancy and recur, regularly, into old age. But none of this was commonly known when I was a child.

These repeated experiences had a radically different character from dreams. From start to finish I would be unable to move or resist. Afterwards I was acutely aware there had been an unwelcome intrusive presence in the house and was left feeling troubled and uneasy. Subsequently, I would sometimes sleep and dream normally, but on waking in the morning would still strongly recall the paralysis, being physically moved and surrounded by 'them'. It was always the same basic routine.

My dread of hypodermic needles and medical procedures was not overcome until, in my twenties, I deliberately steered my career into the healthcare/medical/surgical industry, beginning with a company which manufactured and sold hypodermic needles (of all things) as well as evacuated blood sample bottles in several sizes. These contained anticoagulants for the various types of blood tests carried out in hospital pathology labs. Gradually, with the familiarity of repeated exposure in ways over which I had some personal agency, over time my phobia around medical procedures moderated but it took several years before I could tolerate working in an operating theatre—even just watching the procedure—without keeling over with nausea into unconsciousness.

Aged ten, things took another turn. For several weeks, I had major night terrors. I cowered with my back pressed into the corner of the room in an upright foetal position and was unable to sleep with some non-specific anxiety. If I stretched out in bed to try and sleep normally, I would be terrified of 'grey snakes' in the bed by my feet, grasping me (a deep phobia about the long fingers of the 'little greys'). In the dark it was worse, so my parents consented to leave the hall light on. However, this was almost worse still, as the light from another room shining through the crack in the partly-open door seemed menacing and malevolent, and I was continuously anxious 'they' would come into the room.

My mother tried to be reassuring but I now think was in denial about what was happening. My father confessed in later years that they were seriously worried and for a time considered asking the family doctor to refer me to a child psychiatrist.

The 'Family Line'

It was and is possible to interact with these abductors in a limited fashion, as they communicate directly into your mind and will sometimes respond in a bored-sounding 'voice' if engaged in conversation. All communication with and between these beings is telepathic (yes, it really is) but there is no question if one of them is 'talking' to you directly and responding to your questions as you 'hear' them in your head and the thoughts are definitely coming into your brain from outside in the 'voice' of the individual addressing you.

Some of the beings are more chatty than others, notably the taller ones (they come in taller and shorter varieties, and each type performs a different function). Once, at around ten years old, I remember asking with genuine and respectful curiosity why they were repeatedly doing this to me. The taller one, who was 'in charge', responded matter-of-factly that it was *"because of your mother and grandmother and great-grandmother,"* and as I was next in the family line they "had" to do these procedures, or had some kind of "right" or "need" to do them. Images of these female ancestors when they were young adults were placed in my mind; although I had known my grandmother only as middle-aged and great-grandmother only as an elderly lady who died aged 90 in 1963, I knew for certain that the images of these young women in my mind were genuinely of them.

I could make absolutely no sense of this information about the abductions being intergenerational until forty years later. I do not know why this particularly informative communication was retained in my memory with such clarity, when so much detail of the abductions was effectively blocked. (It seems the abductors may store the event deep in the long-term memory to make it inaccessible, which we'll go into in Chapter Seven.) I still do not understand completely why they work down the generations of the same family line, and I'm not sure anyone (except they, of course) really understands why, but this is a recognised and important facet of the phenomenon and that's what they do. We shall explore some ideas about what might be going on here in Chapter Eight.

My maternal grandmother died suddenly in 1968 at age sixty when I was twelve, but during her life we had a particularly strong bond. Possibly spurred subconsciously by the taller grey's answer to my question, I remember talking to her about the night-time experiences when ten or eleven years old. She referred to the creatures as "the pixies" and reassured me they would not harm me and always return me home. She seemed to understand the experience in some detail, and presumably thought this happened to everyone, or at least to most people, and it just needed to be explained to every child when the child was ready to ask. I only remember one conversation with her about this, but it left a lasting impression and demonstrated that these experiences were not unique to me, and so not just "a product of my imagination."

This conversation with my maternal grandmother when I was around eleven years old confirmed that she knew about the abductions and to a degree understood them, in as much as she knew they would happen repeatedly but felt confident enough to reassure me that they would "always bring you back". I don't even know if she and my mother ever discussed the issue during my mother's childhood (she had a twin brother, a detail which may or may not be important) but suspect not, as so far as I can recall my mother's awareness of what was going on only emerged from her mid-thirties.

It is important to emphasise that, throughout my childhood while these experiences were going on, I had absolutely no clue that they might have anything to do with unidentified flying objects or alien life forms. To me, these night-time visitations were part of the strange 'otherworld': the occult area of folk tales, the dream world, the faery folk, the pixies and all that stuff, disconnected completely from science, cosmology, astrophysics and the possibilities of life on planets elsewhere in the galaxy. These *Magonia*[1] 'explanations' involving temporary abductions by faery folk ('pixies') indeed do serve very effectively to muddy the waters around the subject and ensure, among some people, that the nature of the phenomenon continues to be misunderstood. It is likely that many abductees in many parts of the world during the last century lived their troubled lives embedded in this 'faery world' paradigm, and never even considered that the causative agency might be extraterrestrial.

However, in 1967 I unexpectedly came across something which, in my case, began to radically change that perspective and—slowly—opened the door to new possibilities about what might be going on.

'The Flying Saucer Menace'

One day in late 1967, in a newsagent's shop in the West Yorkshire village (Thornton, a few miles due west of Bradford) where my maternal grandparents lived, I saw a magazine whose title was *The Flying Saucer Menace*. Written by Brad Steiger and published by Universal, its UK distributor was Tandem and it was priced at $3/6$ (three shillings and sixpence in 'old money'). On the cover was a colour photo of a hovering UFO taken somewhere in the USA.

Although I knew at this point (age eleven) next to nothing about the 'flying saucer' or UFO phenomenon, deep down, while staring at the cover of this magazine on the high display shelf in that newsagent's shop, I knew this was important to me and could not avert my eyes from it. Although I couldn't reach it on its high shelf to open it and browse, I had enough understanding even at age eleven to realise that this publication was not a child's comic book but an attempt to understand what the author/publisher considered to be a factual, if mysterious, reality.[2]

Eventually I saved up and bought this soft-covered magazine, read and re-read it with unquenchable obsession and fascination. It was full of accounts of UFO sightings and monochrome photographs from all over the world, a phenomenon of which prior to this date I had almost no knowledge, save maybe for the titles of a couple of 1950s Hollywood movies heard about but not seen. I remember trying (at age eleven!) to persuade everyone that the accounts in the magazine were true and that these must be alien space visitors. No one was much interested, and my mother in particular reacted with anxiety at the very mention of the subject. My father was essentially incurious about the phenomenon and became bored by my persistence in trying to interest him in the 'saucers'. (Ironically, less than one year after these conversations he was the sole witness—so far as I am aware—one dark, pre-dawn morning, to a UFO hovering directly over the neighbourhood.)

Steiger's text contained very brief accounts of four early abduction cases: the Villas-Boas case; the Hills, including reference to their hypnosis sessions with Benjamin Simon, but with no other details;[3] the 1965 Ryerson case in Renton WA; the Valensole case, in the foothills of the French Alps, involving farmer Maurice Masse who described seeing a landed UFO in close proximity, and being confronted by two classic "small grey aliens" who paralysed him. From a more sophisticated 21st century perspective, Steiger's publication does contain a few questionable cases, notably the controversial claims of 'contactees' such as Adamski and Menger. In the 1950s and 1960s, thousands of people thought the contactees were genuine so Steiger probably felt he needed to include them. But the overall thrust of the magazine was one of sober, careful study of a real phenomenon.

I could not understand how people could be so indifferent to something of such importance. However like many unaware abductees, I never connected the dots to conclude these visitors might be intimately involved with me/us. I knew, viscerally, that these UFOs were real, and probably of extraterrestrial origin—though I had as yet to make the connection with my 'pixies'.

The Bleeding Nose Issue

The night terrors did eventually subside and I began sleeping somewhat normally. Like many abductees, I have often been uneasy at night, frequently sleeping until midnight or 1am and then staying awake until around 4am, after which I would again sleep for a few hours. This was a pattern for many years, persisting into middle-age.

I mentioned above that we lived in the small market town of Guisborough, Cleveland—right next to the mediaeval priory that is a regional landmark and iconic symbol of this ancient market town—from 1963 to 1965. This was at the height of The Beatles' meteoric rise to fame, when I was between seven and nine years old. Here in particular I suffered from inexplicable nosebleeds, usually on waking in the morning but more than once at school during the day when writing at my desk. The bleeds were generally from the right nostril, but occasionally from the left and sometimes from both simultaneously. Because my mother had suffered similar nosebleeds as a girl, she thought it was normal. But I can't remember any other kid in the class of 30+ students suffering these nosebleeds during school. In fact, I became somewhat notorious for them; they were often so severe and protracted the school authorities approached my parents about consulting a physician.

We duly visited the family doctor who examined me, but I can't remember much in the way of useful treatment or advice being offered. An x-ray might have revealed something interesting in the top of the nasal passages, but I have no memory of one being ordered or carried out. In any event, the bleedings stopped around age ten and have rarely recurred. (I had one the same year, 2022, as I wrote this book—more on this in Chapter Four.)

It was many decades before I discovered that episodic bleeding from the nostrils, chronic sinus pain and infections, earaches, and persistent headaches are standard stuff for many abductees. They are not universal but are very common. Small implants have been discovered with x-rays and occasionally surgically removed from abductees. Abductees admit to sneezing out small hard objects and obsessively trying to get rid of them by, typically, flushing them down the toilet or washing them down the wash basin drain with copious quantities of water. The precise function of these small implants, of course, remains the subject of speculation but the evidence of their frequent presence continues to grow.

The issue of implants in general, and in particular at the site of the olfactory mucous membrane at the edge of the brain, will be examined in Chapter Four.

Dad's UFO Sighting

In early 1968, my father was up early one morning before 6am to set off on a trip (perhaps a book-buying trip to London or some other place). When he opened the curtains downstairs to reveal what he expected to be the pre-dawn darkness, he saw a huge, lighted disk hovering over the neighbourhood. For reasons he couldn't explain, he didn't wake any of us; he simply got on with preparations for his impending departure and didn't even tell us about his sighting until his return a couple of days later.

None of us connected what he had seen to the night intrusions and other strangeness my mother and I were experiencing. This is one of the anomalies of the phenomenon. It's just possible that it was a random sighting of an unrelated aerial phenomenon, but I doubt it. I myself did not have a clear and remembered sighting of a UFO until the summer of 1972. As far as I am aware, neither my mother nor grandmother ever saw one. Years later, I asked my father about this incident in 1968 and he barely remembered it. But at the time he'd been unequivocal about what he had seen, though curiously reluctant to wake us. Perhaps it hadn't occurred to him to bother us with it.

It must be admitted that, in common with millions of people, my father was throughout his life fundamentally incurious about the extraordinary and the 'strange'. Having been brought up in a fairly strict Methodist (Protestant nonconformist) family and having endured an excess of formal religious education as a child, in adult life he emphatically repudiated his religious upbringing and determined to go his own way. He wasn't the kind of guy who read up on oriental philosophy or studied the *I Ching*, or researched the Tarot, or looked into astrology, or told ghost stories, or was attracted to spiritualism, or even studied western philosophy or scientific advances in cosmology, and he was certainly incurious about the possibilities of alien life elsewhere in the cosmos. He readily recognised that he was married to a woman who was very drawn to "all that strange stuff" and that his son was likewise increasingly interested in it as he grew to maturity, but he equally recognised that this whole area of human experience had no deep attraction for him personally. So, sighting a big, brightly glowing disk hovering silently over the neighbourhood early one morning was something he was simply able to shrug off, and move on with his day. He was a classic example of *The Incurious*, the fourth category in Budd Hopkins' list of reactions to the phenomenon explained so lucidly and succinctly in Chapter Three.

Missing in Norfolk

Aged twelve, the parents of a school friend invited me to go away on holiday with them on a caravan and camping trip to Norfolk (not Norfolk Virginia, but the Norfolk in East Anglia from where Boudicca led the Iceni Tribe in their revolt against the Roman occupation 2,000 years ago).

All went well until, in the middle of the week, the four of us were shopping in Norwich. I was looking in a shop window at some aircraft models in a quiet, almost deserted street when suddenly my friend's parents grabbed onto me very distraught and anxious, demanding to know, "Where have you been?"

I could see the worry and stress on their faces was caused by genuine anxiety. Their behaviour was a mystery to me because I had no idea what they were talking about; I had been with them the whole time and talking with them not a minute earlier and (as far as I was aware) had never moved from the spot.

The three of them told me they had been unable to find me for more than an hour and had searched the area for me, asking people if they had seen a child of my description wandering about or—horror of horrors—whether a child had been seen being abducted by strangers. (Oh, the irony.)

You can understand the anxiety they must have felt in being responsible for someone else's child when he went missing. The prospect of breaking the news of my sudden disappearance to my parents must have been traumatising, not to mention dealing with the police, the negative press and TV reports, and everything else which would inevitably follow, even were the story to have had a happy ending in that I might eventually be found unharmed. I had simply disappeared without trace: there one minute, and seconds later —just gone.

This incident was a genuine mystery to me (and of course to everyone else) as I was sure that I had not moved from the spot in front of the shop window. I could not understand the anxiety that my 'disappearance' had caused. It's less of a mystery now. Yet if it was an abduction—likely, because for a lifelong abductee the evidence offers no credible alternative explanation—it does beg the question: why take me then and there? Did they not realise that I might be missed for an hour, or did they not care? I had been sleeping at night in the caravan with the three of them—did this situation for some reason make abduction more difficult than if I were at home in bed in my parents' house? Normally they 'shut down' potential witnesses into unconsciousness. This practice you would think should be easier if such potential proximity witnesses are asleep anyway.

So many unanswered questions. These parents of my school friend—I'll call him Paul—changed in their attitude to me after this incident, as they could not accept that I had not wandered off for an hour and must have thought me either dishonest or suffering some kind of undiagnosed memory loss psychosis. I assumed that they believed that I had lied to them about not having wandered off for an hour. What else could they think? And how else could they begin to make sense of it?

The invitation to holiday with them was never repeated: a shame, because the following year they went to Madeira, which sounded a lot more exotic than Norfolk.

Apollo 11: Sur la Lune, en Français

In summer of 1969, when I was thirteen, my parents took us to France, towing a caravan. Air conditioning was a rarity in the 1960s found only in high-end luxury cars sold in hot climates, and even then the early technology was cumbersome and did not work very well. Neither the car nor caravan had air con fitted, so we toured France for three weeks through a boiling heatwave with temperatures in the high 30s. We had the car windows open all day on the autoroutes in the vain hope that the scorching hot air passing through the interior from the car's motion might cool us down a bit. It didn't.

On July 20th we stayed for a couple of nights with some of my father's friends in Lyon, so were able to watch Armstrong's and Aldrin's Apollo 11 Moon Lander (the *Eagle*) touch down in the Sea of Tranquillity live on French TV in grainy monochrome, along with the rest of the world. Like most people, I thought that this first Moon landing was the greatest thing ever achieved by the human race; that, as a species we were now bound for the stars, looking forward to a future where all the sci-fi comic-book stories of meeting friendly alien civilizations would soon be realized. Maybe the beings piloting the flying saucers featured throughout *The Flying Saucer Menace* would soon reveal themselves to us and demonstrate their benign intentions to help us reach the stars. At any rate, kids of my generation took it for granted that we would be living on, or at least travelling regularly to, the Moon and Mars within a couple of decades.

To date, the beings controlling the UFOs have declined to reveal their intentions. On the contrary, they often seem determined to go to great lengths to conceal those intentions.

Mutilated Kerry Cow

In summer of 1970 we visited southwest Ireland for two weeks, staying on a farm in Co. Kerry near the coast. I was fourteen. We rode horses and helped to bring the cows in for milking or back to pasture. In contrast to England where almost everything had been motorised since the 1940s, the use of horse and donkey carts as daily transport by the farmers in rural Ireland was routine in those days, so it was an adventure for us in a bygone age of romance we had never experienced.

One day after breakfast, the small son of the family we were staying with, about eleven or so, came to me and said that he needed to show me something. He led the way up a gully between two hills and we walked for about ten minutes away from the farm. We rounded a rocky outcrop and there in front of us was a shocking sight: one of their cows was lying dead on its back, its four legs sticking up rigid at an angle. Part of the udders had been removed, and where the rectum and sex organs had been there was a neatly cored cavity about six inches (15cm) in diameter with a dark red rim around the edge of the hole, as though it had been burned with a hot iron. One of the animal's ears had also been removed, as had part of the mouth tissue revealing the clean white jawbone and teeth, and one of the eyes was missing leaving a neat, dark red hole.

There was no blood anywhere on the ground, no other animal tracks (Ireland has in any case no large natural predators that might kill a cow) and no sign of the missing parts of the cow.

The lad told me the vet had just been to see the carcass, and in his professional judgment the cow had "died of anthrax." The young cow had been perfectly healthy the night before, and the family were upset because she had been a well-behaved animal and a good milker with years of life ahead. One further thing he told me was that for some reason the farm dog would not go near the carcass and reacted with uncharacteristic anxiety when it came near.

Another aspect that was strange (as if it weren't all strange enough) was where the cow was found: it was not in the enclosed pasture with the rest of the herd where it should have been, but part way up this mountain gully half a mile away. It could not possibly have arrived there unless it had either vaulted the fencing—which cows don't do—or been taken there by somebody—or something—so it could then "die of anthrax."

As a result of this experience, I lived my young adult life believing that anthrax was a disease which suddenly struck down healthy animals at random overnight, ate out their insides with surgical precision, one ear, one eye and part of the mouth, whilst leaving no blood anywhere, and affecting no other animal in the herd.

I can no longer recall the exact location where this happened but it was near a small market town where the owner of every store, and almost every resident, was called O'Sullivan. It was only two or three miles from the sea.

One of the curious things about this, to which I never gave much thought until recently, was that the lad took only me up the hill to see the dead cow. My younger sister was also with us at the farm and we hung out together most of the time, but he chose to exclude her from the experience. Perhaps he thought it would upset her, and maybe he thought—or had been told by his parents—not to show the adult guests the dead cow in case they upped and left out of fear of infectious disease risk. (NB anthrax is not a contagious disease but bacillus anthracis, the causative agent of the disease, resides in infected animal carcasses and occasionally in the soil: the spores are very fastidious and can survive inactive for decades without finding a host organism.) At any rate, so far as I am aware only I was chosen among the guests to be shown the dead cow. The other oddity was that the farm had never suffered this before, and the attack/mutilation coincided with our visit.

It was only many years later that I discovered how this precise pattern of cattle mutilation had been observed, recorded and photographed at many places around the world, and that it has often been associated with the sighting of illuminated unidentified flying objects nearby on the nights the animals are killed.[4]

In February 2008 I met Linda Moulton Howe in Eureka Springs AR (and danced with her at a party). She was a TV reporter who had made a hard-hitting documentary film in the 1990s titled, *Alien Harvest*, about identical cattle mutilations all over the American Midwest, and followed up with book of the same title. The very earliest cases, in the early 20th century, were reported in a flock of sheep in Australia, but in September 1967 came the notorious case of a mare called Lady, found mutilated near San Luis, Colorado. By the 1990s these incidents, the huge majority in cattle, had become ubiquitous in the rural states of the US that are dominated by big cattle farms.

All the cattle slaughtered like this looked exactly like that cow in Co. Kerry, but in Ireland in 1970 (and in England for that matter) no one knew anything about this. It may be that I witnessed the first such incident. I have no idea how the carcass was disposed of: if the vet diagnosed that the animal died of anthrax, then county-wide—or even nationwide—measures ought to have been introduced, including the precautionary evacuation of the local population, but as far as I am aware no such action was taken.

The Trigger Event, July 1972

In July 1972 when I was sixteen, the *trigger event* described in the prologue happened on a Saturday morning—actually on 22 July, almost three years to the day after the Apollo 11 Moon landing.

After that morning—with the clear close-up sighting of a flying saucer which I assumed I may have been aboard; the two hours and twenty minutes of 'missing time'; the burn mark on my cheek—I could not deny that some strangeness was intruding into my life. Moreover, that it didn't seem to be random or accidental, but rather, I was beginning to believe that all of this was intentional. Whatever that was supposed to mean.

Of all the weirdness and unconscious suppressed memories that troubled me through my young adult life, this event was the most significant because it was so darned *undeniable*. It was the 'trigger event' experienced by so many abductees, yet as usually happens in these circumstances I did not directly connect it to the repeated night-time encounters with the 'pixies', the prolific nose bleeds and other incidents outlined in this narrative. Even so, the memory block was sufficient to make me almost forget the July 1972 incident for long periods, before it would suddenly come back into conscious memory, one piece at a time. There was always an uneasy feeling about that morning which I could never resolve: I even forgot for long periods of time the clear UFO sighting, so effectively was it submerged. It is this kind of event which leads abductees to believe they were "abducted by aliens that one time when they were sixteen," or nineteen, or twenty-five, or whatever age it happened to them; they usually don't realise that this is not an isolated incident but part of a lifelong pattern, as they have no understanding of how the program works. You're either the one-in-fifty who is in it for life, or else you're not. But people in general don't have even a basic understanding about this phenomenon and how it works. How can they?

Like almost all the minority of partly self-aware abductees at that time, I had no reference points and no idea where to go to begin to understand it. How do you begin to unravel an intrusive mystery that isn't supposed to be real, but which you know is nevertheless happening in your life?

Coda to Trigger Event ...

There was another curious incident when I was seventeen, just a year after the early morning building-site incident. I was travelling on the top deck of a bus with my girlfriend at the time, when out of the window I saw a flying saucer identical to that seen on that summer morning in July 1972. Very distinctive, very clear, it glided along the treetops in the countryside apparently pacing the bus. I felt a strong telepathic connection with it. (I know this sentence reads as rather bizarre, but that summarises the powerful feeling experienced while looking at it.) I excitedly grabbed my companion, pointed out of the window telling her to look—and it vanished. She claimed she saw a "silver blur," nothing else.

Hurst Castle

In 1976 I was at university in Southampton, on the south coast, living in a house with four other students. It was a scorching hot summer, no rain for weeks and the countryside was brown and tinder-dry, with frequent brush fires breaking out all over southern England. Night-time temperatures were up in the high 20s c with daytime temperatures in the 30s and no respite from the heat.

Late one Saturday evening, one of the female students in the house brought back two older friends she knew, both academics and college teachers, from the local pub. They had had the 'spontaneous idea' that they would come fetch me to go with them to Hurst Castle by the sea, some distance to the west of town at the edge of the New Forest. This is an old artillery fortress built in the 1500s to guard the western entry to the Solent, strengthened and expanded over the centuries and in a spectacular location. It's now, in the 21st century, managed by English Heritage. The structure lies at the end of a sand and shingle promontory which juts out into the Solent. There is no road access to the fortress, but an infrequent ferry service will take you there in the daytime. On a hot summer night at midnight, slogging out on foot over the shingle promontory to the castle is the only option.[5]

One of the two college teachers had an old Triumph Herald convertible, so off we went. This incident was slightly odd in that the two guys were not really friends of mine. I had met one of them a couple of times previously, but the driver only on this one occasion and never again. The girl however I knew very well: she was a psychology student at the university, and one of my housemates.

We arrived about midnight and walked out to the castle. It was a beautiful warm starlit night with the waves gently moving the pebbles, and a half-moon visible. We were the only people around. The weather was such during that particular summer that we wore only shorts and T-shirts when outdoors in the middle of the night, and were still very warm.

I remember watching a brightly illuminated, oval-shaped light over the sea as it moved up and down and then stopped. There seemed to then be a missing hour or so. Somehow I found myself lying on my back on top of a low wall maybe two hundred metres away from the other three. The UFO was directly above me, a silent glowing white disk quite high up. I had the distinct impression that it was there because of me and wanted me to look at it. The feeling was strong and positive, like—don't worry, I realise how strange and improbable this sounds—"they" were somehow very pleased with me. The round, bright white object then started to move slowly then stopped; moved slowly in a different direction and stopped; then zig-zagged silently across the sky, and finally seemed to change shape from a disk to a kind of long bar of light and disappeared in a blur into the starlit night.

My three companions remembered only having watched a light in the sky. I alone had experienced missing time and had 'woken up' some two hundred metres away from where I had been what seemed like just a moment before. The others had found themselves sitting on the shingle looking out to sea, assuming that I had wandered off on my own. The most powerful and important thing about the episode for me was not that I had seen another 'flying saucer'—albeit from a greater distance than at the building site four years previously —but the way I felt about it inside. There was a connection of some kind with me and, this time, a very positive feeling.

There was no internet in those days, and anyway we had no camera and it was dark, so no images. And who would you report a sighting to? I had no idea at the time. I told friends, who all said they believed me, and it transpired that some of them had had UFO sightings over the years as well.

One further odd thing happened on our way back to the city in the open-top car, around 2 or 2:30am. A badger walked across the road right in front of the car, forcing us to slow down to a stop to avoid hitting it. It just stood there staring at us with its big, green eyes for a couple of minutes before walking off into the roadside shrubbery.

It was only thirty years later that I realized this episode was almost certainly another abduction. The fact that an hour or so went missing and that I became so physically separated from the other three I always felt to be unresolved and odd but had no idea for years that this conforms to a classic pattern. Perhaps the strangest part of the incident was the compulsion admitted to by the other three that, around midnight, they all had the spontaneous idea that they should come and find me and drive me out to Hurst: one of them I hardly knew, and the driver so necessary to the plan I had never met before nor ever afterwards.

Also at That Time ...

During my 20s, from 1976 to 1986, I had recurring, vivid dreams that I was levitating or flying, complete with the very physical pit-of-the-stomach sensation of dropping swiftly in an elevator, or in an aircraft experiencing turbulence. Some of these included scenes of small children at play. One time, I woke up with a very physical bump, as if I had been dropped back on the bed from a height of a metre or so and landed hard on the mattress. If any sense could be made of these experiences, I tried to rationalise them as some kind of astral travelling experience.

There were many other strange incidents. During one dark December evening in the late 1970s, a spectacle was witnessed by me, accompanied by a second witness, as the sky above her parents' family home in rural Bedfordshire came alive for a while with coloured lights moving slowly around in patterns. They looked like airplanes high up, but each was a different bright colour—blue, green, red, yellow, white. We stood outside in the yard staring at this display, which lasted for around ten minutes.

The other witness (I'll call her 'Florence'), who accompanied me as we watched this display, clearly remembered the event from more than forty years earlier and sent this message to me in July 2022:

> Yes, I remember it clearly, and have often thought back to it. It lasted about ten minutes. The cynic in me says meteor shower; but the fact that they were different colours, and individual lights frequently changed direction and zig-zagged across the sky, at varying speeds, says not meteor shower. I have never decided what they actually were. There were several in the sky at any one time; I think there were probably about 30 or 40 of them all together, over the ten minutes.
>
> I'm glad you remember it too. I would doubt my recollection otherwise. Which is probably what you feel too.

Florence now recalls the year we witnessed this spectacle as Christmas of either 1977 or 1978, but I personally think it may have been Christmas of 1979; we were at this location together in the Christmas season during all those years.

Communion

In early 1987 I was walking down Tooting High Street in South London when I saw a display in the front window of a book shop promoting Whitley Strieber's new book, *Communion*. The face on the cover was arresting and gave me a kind of chill of anxious recognition down the spine. I bought the book, hurried home and read it over the weekend with a combination of fear, anxiety and fascination. It was obvious that the author believed these events had happened as described: truth has a kind of ring to it. For the first time in many years that early-morning incident in 1972 re-emerged with an uneasy feeling.

As a best-selling author whose publisher had a highly effective professional promotion machine, Strieber (with whom I later briefly corresponded) was in

With Tim Good, Beckenham, 2008. Tim and I were both London residents who came to know each other some years after I became acquainted with his books on UFOs.

the UK working the primetime chat-show circuit to promote *Communion*. I was impressed by his careful, serious demeanour when interviewed about the superficially outlandish events related in the book; he would not be dissuaded or cowed by the predictable debunking 'sceptics' inevitably wheeled out to counter his narrative, but seemed genuinely troubled by real, visceral experiences and was seeking to make sense of them.

Many people I knew had also read the book and had been disturbed by it. It is reportedly common for a reader to be unable to endure the alien image on the cover—painted by artist Ted Seth Jacobs from Strieber's description—sitting around the house, staring out. I had exactly this reaction: the book was always put to rest with the image facing downwards to the table. Though abductees report that the entity's head shape is not quite right—the cranium above the eyes is not large enough—the portrait is an arresting likeness capturing the essence of a real 'grey alien'.[6]

The Beginning of a Serious Engagement With the Subject

Over the years I had read intermittently about the UFO subject and was particularly impressed by Timothy Good's *Above Top Secret*. However, I had already assumed the phenomenon was probably ET in origin and didn't need further convincing, so this reading was an educational fill-in as to where the rest of the world was with the subject. I later came to know this author well, and spent time with him in London where we both resided during those years.

Although Strieber's book had opened the door to abduction by these entities and had introduced the pioneering work of Budd Hopkins to a wider international audience (*Missing Time* was published in 1982, and *Intruders* in 1987 concurrently with *Communion*), I still—perversely you might think—did not seriously consider that I might be an abductee. Indeed, my conscious stance on the subject was that I was grateful for that and never wanted to be. The thing is, it's subtle, and hidden beneath layers of memory blocks, so most abductees have only odd, barely-remembered experiences accompanied by a deep, disturbing feeling that something is going on, but one can never quite get to it.

Regardless, I became convinced that alien abduction was real on some level and was probably happening to people, because the narrative pattern across many different cases over time and distance looked to be so similar and so widespread. Even I could see at this time that to explain away these highly complex, precise and often multiply-witnessed events as 'sleep paralysis' was just bunk, of the most ridiculous kind. These things were obviously happening to people, physically real, and caused by some intrusive agency separate to themselves. I was intrigued by the occasionally reported but consistent marks and physical scars on these people's bodies, usually the same marks in the same bodily locations.

The abductors have a way of making you bury and forget the incidents until you start to open up to the possibility, then the floodgates open and suddenly you remember more and it starts to join up.

The Interrupted Journey?

In March 1997 I experienced a very strange car journey from Reading to Amersham, which in this case took around two hours and 45 minutes but should have been less than one hour. There were many odd aspects of this drive: the complete absence of any traffic at what should have been a busy time of day; meeting another vehicle with bright headlights switched on which seemed to be blocking the road; suddenly finding myself in a line of traffic going into Amersham in the dark, when (apparently) a second earlier I had been on a deserted road twenty miles away in daylight with the sun visible in the sky. On finally getting home my shirt buttons were done up wrong.

Had they been like that all day? Unlikely: I had been in a business meeting in the afternoon and surely would have noticed, as would the other attendees.

A cherished piece of jewellery on a heavy silver chain worn under my shirt was missing. I spent weeks looking for it, but to this day it has never turned up. Some other abductee, taken at the same time and in the same location as I was, was perhaps sent home wearing it and no doubt to this day, still wonders how and where it was acquired.

Bosa: a Close Encounter of the Third Kind

In May 2006 during a short business trip to Sardinia, another extraordinary incident occurred. This time, I came face to face with small grey aliens in a hotel room and afterwards remembered the encounter quite clearly.

I was travelling with two others: a business partner from Holland, and my then-girlfriend who came along for the ride so we could spend a few days together in Sardinia afterwards. We spent the first night in a place to the west of Cagliari called Iglesias (English: 'Churches') where one of my Italian suppliers of medical-surgical devices owned an impressive new factory. We then drove by rental car up to Bosa on the northwest coast.[7] We arrived around midnight and, as we needed to get up early to take my colleague to the airport for his flight to Amsterdam, we set my cellphone alarm for 7am.

Now here's the weird part. I woke suddenly, sitting upright on top of the bedclothes with my legs stretched out in front of me. In the room were three figures: two looked like your classic small grey aliens, very thin and with bulbous heads and big, dark almond-shaped eyes. One of them had something in its right hand, like a kind of elongated capital letter 'H' thing about one metre in length (about as tall as he was). Between these two entities was *my father*, who had died three years earlier following a long illness. He 'spoke' to me, saying, "It's time to go, Steve." Then there was a ringing sound like a small bell and the three figures disappeared, leaving only a soft white glow in the room where they had been.

We were a half hour late: it was 7:30, my female companion was in deep sleep and had not woken up.

At the time I thought this might be some kind of visitation from beyond the grave: my Dad coming to say, "Hi" and telling me everything with him was now OK. But the alien-looking entities flanking him disturbed me. Were these strange beings denizens of the 'afterworld' where he now resided? This is how an unaware abductee thinks, and how he tries to make sense of things like this.

Later, I realised that the image was a *screen memory*, not my father at all. They can project these images into your mind, and I have experienced other examples of it. During a night-time abduction in August 2008 I was being moved upwards through the night air, accompanied by what I thought was my then 6-year-old second cousin with her long, wavy blond hair and wearing a long night gown. When I reached out my hand to her, my arm went straight through the image and touched a thin, cold grey alien torso behind 'her'. The creature was taken aback and alarmed, I could feel it: because its 'cover was blown'? The images are very real and technicolour, but in my experience they don't move and the facial expressions (in both instances) were each fixed in a broad, welcoming smile. In each case, the image was projected in front of the alien being (or perhaps projected *by him*), not a detached and separate thing in an otherwise empty space.

Why are these seemingly reassuring or comforting images placed in the minds of abductees? In each of these cases, the only beings present were three 'small greys' whose functions appear to be restricted to initiating the abduction, escorting the abductee around, carrying out a basic routine medical examination and returning the abductee at the conclusion of the process. Might the images of family members be created by the greys to calm and placate the person being abducted, to prevent distress and assist with compliance? This may be considered especially necessary where an abductee has demonstrated a history of resistance or non-co-operation with the abductors. They don't want no trouble, no sir: a smooth operation makes everyone's life easier.

Let me be clear: the two grey 'alien' beings in that hotel room in Bosa were *right in front of me, in the hotel room, in the flesh. I was awake and returned to normal consciousness*. It's almost as though this confrontation was intentional, and they wanted me to see them and retain the memory. It was probably just a slip, a minor error like allowing me to see the 'flying saucer' up close at the end of the July 1972 abduction and, realising their mistake, they distracted me with a screen image of my late father. "It's time to go, Steve," is the normal communication from them at the end of an abduction event.

You might think this would be a clue as to what had repeatedly been happening to me throughout my life, but for some reason the apparition was interpreted as something else. This deliberate deception using screen memories created by the abductors is common with abductions and very often leads the principals to believe all kinds of things, usually glowingly positive, about the deep meaning of their encounters that are simply not true. These instilled memory deceptions, and the neurological abilities of the abductors to skilfully change what the abductee remembers, are the origin of *almost all* the positive 'messages for humanity' abductees believe they are given and are the major water-muddying factor preventing a true understanding of the phenomenon and its purposes.[8]

The Penny Drops

Only at the end of August 2007 did a combination of factors bring the abduction issue into focus and force me to look for answers. I had just returned to the UK from a 10-day business trip to Ankara, Dubai, Isfahan (fabulous place if you ever have the chance to visit) and Tehran, went from LHR Arrivals home for one night, then next morning to Rotterdam on a North Sea ferry to attend a medical diagnostics conference.

While returning from Rotterdam on the ferry across the North Sea, and then driving home from Harwich down the A12, I could not get the subject of alien abduction out of my mind: it simply would not let me go. A relentless need to understand this phenomenon was born, and a final admission that it was probable that I had been personally abducted in 1972 forced me to stare the possibility in the face and take it seriously. This had taken a long time to acknowledge and was frightening and disturbing as the memories began to emerge and coalesce.

Later still did I acknowledge that there had been multiple incidents, a planned and managed program through childhood, and that my mother and grandmother had probably suffered a similar series of experiences. The more I enquired into it, the more I researched the patterns reported by others, I began to perceive something approaching an understanding, or at least an acknowledgment, of some phenomenon, yet unexplained and mysterious, that had been haunting my life.

Like most abductees when they begin to suspect what has been going on with them, I had nowhere to turn and no one who might be trusted to maintain confidence about the matter. I cautiously reached out to several people with little result. Eventually, I found Nick Pope,[9] who promptly responded to my query and was most helpful. He referred me, with the usual cautions and caveats about hypnosis, to a hypnotherapist in North London who was qualified, registered, and worked with the National Health Service. This hypnotist had worked with several abductees over the years and had a very positive spin on the whole thing but, I later realised, had almost no understanding of the phenomenon's fundamentals or purposes. As I was to learn, this matters if you want to truly

With Nick in London, January 2011.

understand what is going on and not go down one blind alley after another. During two regression sessions with this professional I did not remember very much. (But one medium-term consequence of these sessions is recounted in Chapter Seven, when discussing the delayed effects of hypnosis.)

Nick and I subsequently met up in London and at other locations several times, including at Rendlesham Forest in Suffolk, and at UFO conferences in Glastonbury, Leeds, and Catalonia, and other places. He was above all the most helpful and level-headed individual I encountered during these times. He's also great company. One of his books, *The Uninvited*, examines the abduction cases he investigated while working for the UK Ministry of Defence. It is a great primer, thoughtfully written with intelligence and humour.

Once the door was opened I read the published works of Budd Hopkins, David Jacobs, Ray Fowler, and John Mack, and others. I travelled to the USA many times specifically to pursue this issue, and met with several abductees—Travis Walton, Randall Nickerson, Linda Napolitano, and many others—and got to know many published authors and researchers who have been very supportive. In early 2009 I started working with Dr David Jacobs (now retired from the field after 32 years' engagement with it) and our sessions together were the most helpful in enabling me to see what had been going on. David devoted a great deal of time to abductees once he agreed to work with them, but the contact must always be initiated and maintained by the abductee. He charged nothing for all this care, attention, and time. Later in this book,

Chapter Seven is devoted to detailing my experience working with and getting to know David Jacobs: of all the people I have ever known, he was the one who understood most thoroughly what had been happening to me and his intelligence, thoroughness and intellectual rigour helped me more than anyone else to make sense of it.

Post-Awareness Incidents

This is a short selection of incidents experienced since becoming aware that abduction was the source of all the strangeness in my life. I was 51 years old before I became convinced, so tend to look at these later incidents from a different and slightly more informed perspective than the catalogue of earlier events.

Two In Quick Succession, One With Unforeseen Consequences

In August 2008, I suffered separate abductions on two consecutive nights and vividly remembered both. During one of these episodes, a scoop-biopsy was taken from the right shoulder. The long-term consequences of this event are detailed in Chapter Four, as they led to me being invited to New York City to stay with Budd Hopkins and subsequently to meeting other people prominent in the field, including David Jacobs and Leslie Kean.

Disturbing Weirdness and Returning Memories

Some disturbing memories returned in the middle of the night several times during the latter part of 2008 and into 2009. Often Jan, my partner, would wake me up because of my subdued (and sometimes not so subdued) shouting. The dreams which initiate these episodes are always about UFOs (in one case a very large and distinctive orange one, not in itself frightening but for some reason the initiator of fear) or about being handled in the dark by stealthy and quick-moving small grey creatures. I don't remember any other time in my life when this kind of thing happened. It's like part-memories are returning, with a fear and anxiety present every time. One time I saw a blue-skinned alien—a classic skinny one with bulbous head—at the foot of the bed and was apparently shaken awake while screaming, "Get away! Get away!" This entity had large, vivid blue eyes with spectral blue irises, not the more commonly reported black almond-shaped eyes but was in other respects similar to the classic morphology of the grey alien.

New Mexico and Arizona:
Robert Hastings, Travis Walton and Reuben Crow

In the early summer of 2010, Jan and I attended a surgical conference in Minneapolis MN. When the conference was over, we flew down to Albuquerque in New Mexico and rented a car to tour the Four Corners States, as she is interested in Native American culture, traditions, and art. On this trip we met, and for the first time stayed with, Robert Hastings,[10] with whom I had struck up an online correspondence and gotten to know over the preceding months. Robert is now retired in a very remote location in Colorado, but in 2010 he was living in New Mexico near Taos.[11]

Robert showed us around New Mexico, including some ancient sites such as Chaco Canyon and others in the stunning, arid landscape of the American southwest. After three memorable days, we headed off to Arizona to see the Grand Canyon and try to meet up with the elusive Travis Walton, whose multiply-witnessed five-day UFO abduction near Snowflake AZ in November 1975 made him a reluctant international talk-show celebrity. Travis' story is recounted in Chapter Two.

Prior to meeting Travis, we had stopped to pick up a hitchhiker, a Navajo Indian named Reuben Crow, outside a small town on a reservation in Arizona. From the back seat he initiated a truly extraordinary and memorable conversation, which lasted a couple of hours. With no prompting from us, Reuben introduced the subject of UFOs and the local Navajos' experiences with visiting extraterrestrials. It turned out that everyone in the local area had seen 'flying saucers'—including Reuben, who pointed out through the car window exactly where in the mountain landscape he and others had had a daylight sighting the previous week. It seemed the phenomenon was normalised in the local culture, and his attitude to the issue was very matter-of-fact, as though it were an everyday matter. This episode was one of those conversations that you remember for the rest of your life and talk about over the dinner table.

The Grand Canyon is beyond superlatives. The sheer scale of it is like nothing else you'll ever see. If you get the chance, see if from the air: you can take a helicopter ride, or a light airplane which will fly into the canyon and stand on one wingtip so you can look straight down out of the window. It's not a place for those who suffer from vertigo, or who are even a just little nervous about precipitous heights.

Monument Valley in Utah is also spectacular, with its giant, unworldly sandstone buttes. It has been the location for hundreds of classic Hollywood

Robert Hastings in Hastings, July 2014.

Western films. It's also managed by the Navajo, who run the cafe and gift shop complex. During our visit we experienced a sandstorm so ferocious that it filled the rental car interior with sand, despite closing down the ventilation system, and we had to park up by the roadside as visibility proved impossible beyond the hood of the car. I had visions of our vehicle being buried in the sand as dramatized in Anthony Minghella's 1996 film, *The English Patient*, but fortunately the sandstorm did not endure long enough to necessitate our rescue and we continued on our way.

The Normandy Landing?

In September 2014, our by-then good friend Robert Hastings visited and stayed with us at our Hertfordshire home in Southeast England. He flew from Denver to Heathrow, from where we collected him. After a few days settling in and catching up, the three of us went to stay in Normandy, packing up my station wagon with our luggage and taking the cross-channel ferry from Newhaven to Dieppe, sunbathing on the deck in glorious late-summer sunshine.

Our favourite place to stay in Normandy is Arromanches-les-Bains, the site of Gold Beach from the 1944 D-DAY landings, and where the remains of one of the Mulberry Harbours still lie in the offshore tides. Arromanches offers convenient access to everything of interest to visitors including Caen; Bayeux with its famous mediaeval tapestry describing in 11th century comic-strip format the conquest of Anglo-Saxon England by the invading army of Guillaume-le-Batard; the American war cemetery and memorial overlooking Omaha Beach; and the spectacular, historic Mont St. Michel, one of our favourite locations despite the summer crowds. However, our usual hotel in Arromanches (right on the seafront) had just closed the previous weekend for a major renovation and everything else was fully booked, so on this occasion we had booked to stay in a picturesque farm cottage near Falaise. From Dieppe we drove down to this place and arrived there in the early evening.

Robert takes up the story of what happened there:

We left Hertfordshire early on the morning of September 6, 2014, arriving at the holiday cottage in Normandy shortly before sunset. We didn't even unpack but immediately opened a bottle of wine and went out on the patio. No more than five or ten minutes had passed before I saw a white disc emerging from behind and above a group of trees south of our position and flying directly toward us. It was initially the apparent size of an American quarter-dollar coin held at arm's length. However, as it approached us, it suddenly expanded in size—much more than its forward motion would have generated—and then quickly shrunk back down to nearly its original size as it continued to approach us. Then it abruptly stopped directly over our heads.

Although I am confident that all of the above is completely accurate, once the disc began to hover, I have no clear memory of it leaving. For years I thought that it had blinked out—that is, became invisible in place, as opposed to flying off. But recently I have begun to doubt that. Given the fact that it appeared shortly after we arrived, after traveling all day, I am convinced that those presumably aboard the disc knew of our arrival and had come specifically to interact with us in some manner.

After I watched it hover for a few seconds, I turned to Steve—who was sitting to my right—to see whether he was also watching the spectacle. (Jan had her back to the approaching object and apparently didn't see it.) Steve immediately asked me, "Did you see that?!" All of this seemed to occur only seconds after the object stopped and hovered but now I wonder whether we instead had a missing time experience that commenced just as the hovering began. In fact, although I recall a pink glow in the sky when the disc appeared—suggesting that the sun had not yet set, or perhaps had just set—when I looked over at Steve, the sky seemed darker, as if dusk had settled in. I have recently run the experience through my mind over and over as I fall asleep, trying to visualize the events, the amount of light in the sky, and so on. Yes, the human memory is fallible, so I can't claim that I have accurately remembered the subtleties of the incident, but I do think that it was suddenly darker when Steve spoke to me.

But given that the disc was fairly close, and stopped directly above us, I doubt that this was merely some random sighting. I think they came to take one or more of us. Because of certain events the following morning—including Jan telling me and Steve not to stand too close to her—I think she was subconsciously remembering an intrusive, threatening confrontation of some kind the previous night. Moreover, Steve and I had experienced anomalous sleep patterns, which we mentioned to each other in the morning.

Robert

I confess that my memories are no clearer than Robert's so can add little extra detail to his account, save confirming that a glowing white disk was seen above our heads as we sat at the garden table drinking wine. Then suddenly it was darker, we all realised we were tired and went to bed. I will acknowledge however that we had disturbed, troubled sleep not usually experienced after a long day's travelling. This incident has all the hallmarks of a combined abduction, but of the missing part of the evening none of us has any memory. The jury is still out on this one.

Disturbing Memory

In August 2018 we bought a 140-year-old barn conversion in Lincolnshire, an impulse buy in a location we didn't even know and had never considered. But we fell in love with the house, its isolated location, huge garden and interior space. The location is designated as an Area of Outstanding Natural Beauty (AONB) so new building developments are outlawed unless they replace an existing structure of identical size, and are always subjected to a protracted planning approval process. These onerous rules are designed to safeguard the preservation of the original landscape and make sure it is not despoiled by inappropriate building developments, and ensure the existing housing retains the character and charm of the historic area.

Throughout the winter I travelled up from Hertfordshire for a few days each week to oversee the building works, as the house needed a lot of renovation, and we finally moved our furniture from Hertfordshire to the part-completed, but now habitable, barn conversion on 1st March 2019.

On or about 14th March, we were at the beginning of a journey from our new home to Hoylake on The Wirral to view a litter of pedigree Rough Collie puppies which had been born there on 10th February. Our journey passed a quiet place surrounded by a large open heathland, bordered on all sides by a pine forest. This was the first time we had ever passed this area—or so I at first thought.

A chill of recognition shot through me as we drove past this place. I had a sudden clear memory of having been there at night under a starlit sky, standing in the heathland with around 30 or 40 other people, with a large glowing cigar-shaped object stationary in the sky over the woodland to the north of the main road. This was a searing and visceral memory, totally real. It had happened sometime in the previous six months, as prior to that we had not been living in the area.

Because I had no further detailed memory than this, and no precise date nor further context, I found it difficult to recall the events leading up to or subsequent to the memory of standing in the open area at night together with the 30–40 people looking up at the bright object in the sky which, I recall, filled me with a sense of dread.

A couple of days after we returned from the trip to Hoylake, I went out to the location on a Sunday morning—Sunday 17th March to be precise, St. Patrick's Day. The sense of familiarity and the memory of standing there under the open night sky was even stronger. I also had a powerful memory of having been driven to the location as a back-seat passenger in a car—a VW Golf with a black interior trim to be exact—however improbable that may be. Normal memory can be notoriously unreliable when dealing with this phenomenon, but this one I trust, as the memory of the huge dull-glowing cigar-shaped craft hovering over the woods is strong and unshakeable.

I don't know what to do with this one. Dave Jacobs has stopped working with abductees and retired, so even though our long-distance friendship continues, I am not going to ask him to help me further with this. The event was almost certainly something important or significant to the abductors, for which it's safe to assume they needed thirty or forty abductees together—presumably aboard the huge cigar-shaped craft.

No further information available at the present time.

We bought one of the Collie pups, btw. He still lives with us; he's now three years old and I walk him every day through the local woods and fields.

Further Bodily Interference and Minor Assaults

One night in late October 2021, I awoke around 1:30am with a sense that my right thumb had a damp area on the metacarpal, so got up and went into the kitchen to see in better light what was wrong, without waking or causing alarm to my wife.

The outside surface of the metacarpal section had three equilaterally spaced fresh puncture marks around 1mm apart with fresh blood oozing from each small incision. After staunching the blood flow, around 15 minutes later a scab formed and although the marks were so close together, they could still just about be seen as separate incisions. I wish that I had had the presence of mind to photograph the injury when the three incisions were still fresh, as once the scab formed their symmetry was more difficult to see.

With no memory of an abduction event, the perfectly symmetrical triangular incisions are the only evidence that anything untoward happened that night.

The Nasal Implant

Chapter Four is devoted to an exploration of 'Scoop Marks, bodily scars, implants and forensic analysis of hair'. A section titled, 'My own personal nasal implant "misfortune"' describes my ejecting a nasal implant on Friday, 25 February 2022, which caused bleeding for 24 hours from both nostrils.[12]

Roger Leir—a California surgeon who, between 1995 and 2014, removed 17 implants from abductees—together with Steve Colbern, a chemical materials analyst who has worked on these nano-devices removed from abductees, and others have identified that these implanted devices emit radio signals in specific frequency bands. Moreover, after a few weeks outside the human body they stop transmitting.[13] This design feature is probably engineered-in to alert the abductors that the device has been removed or ejected and now needs to be replaced.

On 3–4 April I believe 'our little friends' replaced the implant, described in 'Postscript to the Incident' in Chapter Four. The newly-sited implant again caused my nose to bleed for more than 24 hours before 'settling down'.

My Mother

My mother was the lifelong abductee in the intergenerational chain; her mother before her, and her mother before her were almost certainly earlier unwilling inductees into the program.[14]

The pattern of my late mother's life makes more sense when seen through the lens of the abduction phenomenon than it did at any time when she was alive. Until her death in September 2000, aged sixty-five, although she did not fully understand what had happened to her throughout her life she felt somehow 'special' and 'watched over' by some agency. To her it was "spiritual" and helped to open her divinatory and clairvoyant faculties. This was possibly a consequence of the abductors' repeated *mindscan* procedures of neurological manipulation, and frequent practice of telepathic communication with them throughout a lifetime. In her 30s she was attracted to seances and spiritualism, always reading books on the subject, and had joined the Spiritualist Association of Great Britain,[15] which was based in London. In my later teen years, she quite often took me out for a Sunday visit to some clairvoyant or crystal ball mystic, on the pretext of me practicing my learner driving skills on the journey. In her 50s she became a successful and reputed spiritual healer and clairvoyant, though was never very easy with it all, and never charged anything to anyone she treated.

Juxtaposed with this were severe gynaecological problems in her 20s and 30s. She had absolutely no idea that any of this might be related to the UFO/ET phenomenon, and of course neither did the rest of us. It never crossed anyone's mind; why should it?

My mother, and her mother, were undoubtedly abducted repeatedly throughout their lives. From what I know now about the nature of this phenomenon, it seems certain that by this intergenerational route I, in turn, became involved with these entities—as one of the abducting entities explained to me in 1966.

She told me repeatedly during the last few years of her life (this would be in the 1990s, as she died in September 2000) that she knew she, and I, and her long-deceased mother, were "involved" with something occult, abnormal but intrusive and physically real. She used to refer to the abductors as *they* or *them*, as she was familiar with them and their often bizarre activities. She blamed them for the many things that would often go missing around the house—items of clothing, bed linen and towels, cutlery and the like—though these would normally be returned within a few days. This made no sense at all to us in the 1970s and 1980s but does fit closely into the narrative of hybrid integration into society revealed during this present century. I have also experienced the missing-items-soon-returned phenomenon, and for years never knew what to make of it except, "it's one of the weird things which happens to me." Abductees Betty Andreasson and Bob Luca also describe this aspect of the phenomenon in one of their books.[16]

In the last two years of her life my mother channelled poetry. She would wake in the middle of the night, or be interrupted from normal daytime activities, and would have to start scribbling down this stuff coming through her mind. There were about 100 poems in all, on a variety of subjects, many quite serious and philosophical. The styles were all different, and some were written in heavy regional dialects (like Robbie Burns' original material for instance) of which she had little or no knowledge. This process carried on regularly for about two years, then just stopped. Quite a large body of work survives. This is probably not *directly* a result of the abductions but, again, the opening of these faculties is possibly a by-product of abduction activity—neural manipulation, intentional memory blocking, frequent telepathic communication with the brain, frequent levitations and so forth. Speculation of course, but *something* is going on here. All my life I have found myself at the least connected to it, and often immersed within it.

Now it's time to look at some classic cases of abduction which found their way into the public domain in the 1950s to 1970s, and some traction started to be gained in the popular consciousness that this phenomenon might indeed be real.

Chapter Two

A Spectrum of Early Cases

Which came to widespread public attention and collectively confirmed what is going on

The cases summarised below have been chosen as examples because either some or all of these factors apply to them.

1. They were early cases prior to 1980, when the UFO abduction issue was not in the public domain to any appreciable degree or openly acknowledged to exist, even among those taking sighting reports seriously. The exception is the Debbie Jordan-Kauble case, which only came to public attention from 1983 but whose primary events date from the 1960s

2. In almost all cases, the witnesses reported a missing time element

3. They were initiated or accompanied by a clear and close-up sighting of an unidentified flying object seen by multiple witnesses, so the direct connection between the missing time, the remembered or later-recalled abduction event, and the unidentified flying object is certain

4. They achieved some national or in many cases international news media exposure. In at least three cases—the Hill case, the Walton case, and the Jordan case—feature films were made about them, each with a cinematic release [1]

5. In most cases the abductions had included multiple participants or had multiple witnesses, reinforcing their likely veracity

6. The incidents were extensively investigated by multiple third-party investigators and well documented, including the publication of books about each case

7. In most cases some professional memory retrieval expertise (sometimes called 'hypnosis' but this term is not always accurate) was deployed at the request of the participants/witnesses to assist in recalling the events in question

8. Not one of these cases has ever been convincingly 'debunked', despite the best efforts of the True Believers (see Chapter Three)

The Cases are, in Chronological Order:

1. Antônio Villas-Boas – Brazil, October 1957
2. Barney and Betty Hill – White Mountains NH, September 1961
3. Betty Andreasson – Ashburnham MA, January 1967
4. Charles Hickson and Calvin Parker – Pascagoula MS, October 1973
5. Travis Walton – Snowflake AZ, November 1975
6. The Allagash Four – Allagash Waterway ME, August 1976
7. The Debbie Jordan-Kauble case – Indianapolis IN, eventually reported in June 1983

Antônio Villas-Boas – October 1957

Antônio Villas-Boas was twenty-three years old when he reported that he had been abducted by strange creatures from a flying saucer near São Francisco de Sales, in far Western Brazil on 16th October, 1957.

Villas-Boas reported that he and his brother had seen a "bright light dancing in the sky," ten days earlier, on October 5th, and again on the night of October 14–15, as they were working the fields of their father's farm at night to avoid the daytime heat.

On the night of the 16th, Antônio was working the same fields alone when he again saw the bright "reddish" light in the sky and subsequently watched it grow in size as it came closer. Then the actual shape of the object came into focus as a metallic, "egg-shaped" craft which landed near him on three legs.

He claimed that creatures about 1.2 metres tall appeared and that he was taken from his tractor by and compelled to enter the craft with them. The only features Villas-Boas reported as memorable about their appearance was that they were thin, hairless, and wore "enormous dark sunglasses" over their eyes (almost certainly referring to the large almond-shaped eyes of the creatures). He also claimed that each was wearing a kind of one-piece body suit.

Villas-Boas described being stripped naked and smeared all over with an odourless liquid, then left alone in a room inside the craft. He said that a gas of some type was then pumped into the room, and that he vomited.

The central memorable detail of the encounter, and that which subsequently made it so controversial, was that a naked woman then entered the room and that Villas-Boas was twice encouraged to copulate with her. He gave a detailed and precise description of her, that she was about four feet six inches (137cm) in height, "with bleach blond hair, big slanted blue eyes, no make-up, straight nose, narrow face, very thin lips almost invisible, soft skin which felt like there was no bone." The colour of her pubic and underarm hair he described (with some embarrassment) as "bright red." According to Villas-Boas, the alien female reportedly bit his chin continuously throughout their encounter but would not kiss. She did not speak, but barked and growled—"making animal grunts"—during their performance. As they separated, she pointed to her abdomen and then upwards towards the sky before leaving.

Despite, or perhaps because of, these lurid details, the case received little publicity outside Brazil until the mid-1960s. The earliest reference to the story appeared in a Brazilian UFO periodical in 1962.[2] The story was not reported in print outside Brazil until 1966, coincident with the publication of John Fuller's book, *The Interrupted Journey*, which investigated the Betty and Barney Hill case. Regardless, most researchers found the Villas-Boas case too embarrassing and outrageous to deal with and so gave it little attention.

Villas-Boas estimated the abduction lasted a total of four hours and fifteen minutes, after which his clothes were returned and he was reunited with his tractor in the field, which he mysteriously discovered had had the battery cables disconnected. (It's difficult to make any sense of this unless his abductors had been skilled automotive mechanics with a convenient box of tools, but it was hardly any easier in 1957 to make sense of the rest of the story.)

There had been similar accounts of abductions in Brazil prior to this, but the Villas Boas case was eventually to gain more attention than most due to several factors:

1. Although originally described as an "illiterate Brazilian peasant" in international press accounts, Villas-Boas was in fact highly educated and became a successful practising lawyer

2. The details of Villas-Boas' story remained consistent throughout his life and he insisted on its veracity until his death in January 1991

3. Elements of the story, though at the time seemed unbelievable and even acutely embarrassing, were to closely correspond in their essential details (the alien breeding program, alien-human hybrids, the specific appearance of the 'alien' female and the 'small greys') with much later accounts of abduction in other parts of the world

4. For three months following the encounter, Villas-Boas suffered moderately severe medical ailments indicative of radiation poisoning: body pains; nausea; headaches; loss of appetite; excessive sleepiness; a constant burning sensation in his eyes; and cutaneous lesions when incurring the slightest bruising, that went on reappearing for months and looked like small red bumps, harder than the surrounding skin and painful when touched, each with a small central orifice emitting a discharge of yellow pus

5. Dr Olavo T. Fontes, Professor of Medicine at the National School of Medicine in Brazil, who treated Villas-Boas' multiple health issues following his experience, persuaded him to go public. Villas-Boas himself was reluctant to do this, as he felt he would not be believed. At Dr Fontes' prompting the story was released to journalist Joas Martins (who was also a Brazilian military intelligence officer) on 22 February 1958, though it initially received no reporting outside Brazil and did not even appear in print inside Brazil until April 1962

Antônio Villas-Boas went on to a successful career as a practising lawyer. He shunned publicity following his 1957 encounter and neither sought nor made any money from it. He was repeatedly invited onto chat shows and approached by newspapers and magazines for his story, but always declined to be further interviewed, stating that he had nothing further to add to the story. He and his wife had four children and he died in January 1991.

Main Takeaways

1. Outdoor abduction with the abductee driving a tractor engaged in farm work. Obviously, the 'sleep paralysis' nonsense is not going to fool anyone here

2. An early—maybe the first—suggestion of a hybrid breeding program, though it is possible that the abductee's experience was subject to some confabulation or memory manipulation

3. Moderately severe negative health consequences for the abductee requiring medical attention for 90 days following the event

4. Early example of 'grey aliens' controlling or directing a hybrid

Barney and Betty Hill – September 1961

The Hills were a mixed-race couple who lived in Portsmouth NH. Barney (1922–1969) was employed by the US Postal Service, while Betty (*née* Eunice Barrett,1919–2004) was a social worker specialising in child protection cases. The couple were active members of the Unitarian congregation, pillars of the community, and committed to the civil rights movement.

On the night of 19–20 September 1961, the Hills were on a late-night drive home to Portsmouth NH from a brief three-day holiday in Montreal and Niagara Falls. In the car were Barney, Betty, and Betty's pet Dachshund, Delsey. The Hills determined they would drive through the night, expecting to arrive home by 2am or 3am, hoping to be a few hours ahead of a hurricane that was forecast to blow in from the Atlantic around dawn on the 20th.

According to the Hills, an object was first sighted in the sky at about 10:30pm, near Lancaster NH, on the edge of the White Mountain Forest. Betty observed a bright light low in the sky, which moved from below the Moon and the planet Jupiter on an upward trajectory to the west of the Moon. The light moved erratically and grew larger and brighter, so Betty urged Barney to stop the car for a closer look. She took this opportunity to walk the Dachshund as it was still several hours' drive to home in Portsmouth.

Betty could see the airborne object flashing multi-coloured lights. Through binoculars, it resolved itself as a large disk-shaped craft, with red lights that seemed, from their perspective, to be attached to protuberances sticking out from the sides of the object. Her sister had several years earlier reported seeing a 'flying saucer' and the phenomenon had been much in the news in the fourteen years since 1947, so Betty thought that this might be what they were seeing.[3] Twenty miles further on their journey, near the resort of Indian Head, the craft descended and blocked the road, and the Hills reported that several figures were clearly visible through the windows. At this point Barney got out of the car and walked towards the craft with his gun but came running back shouting, "They are going to capture us!" He later claimed that one of the occupants had communicated with him telepathically and told him to return to the car and wait. Their memories became murky after this, neither recalling much about what came next until they were some 35 miles south of Indian Head. They arrived home after 5am, two hours later than anticipated and with a sense of unease. They later found that the car trunk had some unexplained shiny round marks, and when a compass was placed near them it unaccountably spun round and round. Barney's shoes were heavily scuffed *on the tops of the toes,* as though he had been dragged over rough ground with his feet dangling. Betty's dress was inexplicably torn and had some unusual pink stains. They both remembered hearing a "beeping" sound coinciding with the UFO sighting at two different times during the journey, between which they remembered nothing but a dream-like haze.

Betty's sister convinced her to contact Pease Air Force Base. During a returned call the next day, according to Betty, a major informed her the base radar had picked up something in that area shortly after 2 am that morning. The case was shortly afterward forwarded to Project Blue Book.

Persistent nightmares and the emergence of what we might now term *post-traumatic stress* symptoms led the Hills eventually to seek medical help. The therapeutic hypnotist who treated them, Dr Benjamin Simon, was a respected physician in Boston who had successfully treated ex-servicemen for combat trauma. He carried out regressive hypnosis sessions with the couple between January and June 1964. Dr Simon had no interest in the UFO phenomenon, nor any knowledge of 'alien abduction'. (There had been no reported cases in the public domain at that time.) The couple were regressed separately by Dr Simon and carefully managed so that they could not communicate their memories to each other. Nonetheless, he was astounded to uncover mutually corroborative and supporting accounts of a complex abduction by apparent extraterrestrials. The Hills themselves, when they heard their taped interviews with Dr Simon, had a hard time coming to terms with what they had related and initially refused to believe their own memories. Dr Simon remained sceptical, though he could not shake the recovered memories and in fact these memories explained all of the physical evidence on their persons.

The Hills separately recounted being taken aboard the craft by strange creatures, and separately given medical examinations before being returned to their car. Betty reported that a needle was inserted into the area below her navel and was told this was a "pregnancy test." The couple separately reported having their eyes, ears, and inside the mouth examined. Their captors were reportedly puzzled as to why Barney had detachable teeth and Betty did not: he was fitted with dentures.[4] Their joints were manipulated, skin scrapings and nail clippings taken for analysis.

There is controversy about a "star map" shown to Betty by the "leader" (almost certainly a grey alien of the taller type) that she recalled in some detail during the hypnosis session. Later, an amateur astronomer named Marjorie Fish identified the stars as part of the Zeta Reticuli binary star system but *not as viewed from the perspective of our solar system.* The existence of this binary star system was unknown in the early 1960s.

The encounter was investigated by the local US Air Force base, and by other organisations, including the National Investigatory Committee on Aerial Phenomena (NICAP). The story initially broke into the public domain in 1965 in an article penned by *Boston Globe* investigative reporter John Luttrell. The next year the encounter was recounted in more detail by John Fuller in his book, *The Interrupted Journey*. From there the incident was rather sensationalised in the national press. The Hills became reluctant celebrities, and this unsought fame brought them many problems. As a mixed-race couple during the sixties, they'd have preferred a much more low-key life. Indeed, they reportedly used to be teased good-naturedly by their friends and fellow Congregationalists because they had the same first names as Barney and Betty Rubble, the Flintstones' neighbours and friends in the popular TV cartoon series of the time. The last thing they wanted, or needed, was to court the kind of negative publicity which might reasonably be expected to accompany such an outlandish and improbable tale in 1961 as being abducted by aliens.

Barney died of a cerebral haemorrhage in 1969, aged 46. Betty survived Barney by 35 years and never remarried; subsequent paranormal incidents followed on the heels of the landmark White Mountains abduction which are detailed in *The Interrupted Journey*, and described in greater detail in *Captured*, authored by Kathy Marden (Betty's niece) and Stanton Friedman.[5]

One important footnote: During his examination aboard the craft Barney had a sperm sample taken, vividly recalled in one of Dr Simon's recorded hypnosis sessions, but Barney was so embarrassed—remember that this was 1961 and the Hills were respectable church-going people—that he later insisted John Fuller *remove all reference to this in his manuscript*. This was not publicly known until decades later and was extremely fortuitous in one respect: when male abductees during the 1970s and '80s started to recall that they had sperm samples taken during their abductions,[6] nobody could claim they had been influenced by the Hill case, as this detail had been excluded from the record and never reported anywhere until finally revealed by Betty Hill in the 1990s.

Main Takeaways

1. Outdoor abduction with two abductees as driver and passenger in a motor vehicle on the highway so again, the 'sleep paralysis' nonsense is obviously going to make no headway here

2. Abduction followed repeated sightings of the UFO, the description of which was later corroborated by other witnesses

3. The first reported instance to come to public attention of the missing time element following a close encounter with an unidentified flying object

4. Perhaps the first example of the use of professional hypnosis to recover buried memories of events which occurred during the missing time period, by a skilled practitioner who had no knowledge of and little interest in the UFO phenomenon and did not initially believe the couple's recollections

5. Details of medical examinations of the two abductees later revealed to include the classic sperm/ova harvesting procedures closely correspond in detail to later accounts from other abductees

6. Much physical evidence (heavily scuffed shoes, torn dress, damaged automobile, magnetic compass anomalies) and PTSD symptoms experienced by witnesses and attested to by investigators and medical personnel

7. Perhaps the single most investigated and written-about incident ever in the history of the abduction phenomenon

Betty Andreasson – January 1967, and afterwards

One reason the Andreasson case is important to the study of the abduction issue is that, when the case was first investigated in the early 1970s, the patterns later uncovered by other independent researchers were unknown and much of this information was new. Some of the details described by Betty Andreasson still seem unique to this case, though increasing knowledge about the phenomenon in later years meant that when seen retrospectively many aspects are now seen to fit a common pattern.

On 25th January 1967, Betty Andreasson's isolated house in South Ashburnham, Massachusetts suffered a mysterious power failure. During the evening she was reportedly abducted from the living room by five small, hairless, and large-eyed 'grey aliens' who walked through the door (i.e. without opening it) while the rest of her extended family, with the exception of her eldest daughter Becky, were rendered temporarily unconscious for the duration. It is not well understood how or why Becky remained fully conscious and subsequently recalled most of what happened inside the house, but she too was later abducted on a separate occasion.

The report of Betty's abduction qualifies as 'high strangeness' by any standards, and at the time seemed unique and extraordinary. It was many years before news of the incident reached investigators because, as the author and researcher Raymond Fowler asked: "Where does someone go to report a UFO experience so bizarre that one hesitates to discuss it with either family or friends? Where does one turn when government officials have publicly decreed that UFOs do not exist? Such was the plight of the Andreasson family."

Eventually a letter from Betty to J. Allen Hynek at Northwestern University found its way to Raymond Fowler and a local Mutual UFO Network (MUFON) investigative team, who thought it might be worth looking into. The case is still one of the best documented ever because of the large quantity of recorded testimony, extensive polygraph testing of the key witnesses, detailed analysis of corroborative evidence, careful character checks, and comparison with other cases. The eventual 528-page report, signed by five investigators, was initially distilled down to a 200-page book authored by Fowler and published in 1979 as *The Andreasson Affair*. Betty Andreasson was a competent artist and the case is also unusual because she was able to render her memories of the aliens and their technology into detailed drawings which, reproduced in full in Fowler's book, add much to his narrative.

The story is problematic to some investigators because Betty, a committed Christian, saw her experiences through the prism of religious belief and of overriding importance to her is to know whether the alien entities are "angels" of God or "demons" of the Devil. In recent years it has been posited that, in order to ensure better compliance, the abductors tend to play to the belief-system and values of the abductee to make the experience more acceptable, and they convinced Andreasson she was having a positive, God-inspired encounter.

Ray Fowler wrote four follow-up books on the Andreasson case,[7] and it transpired that Betty's experience was in fact more similar to others than was evident when the story originally came to light, in that she turned out like most abductees to have had serial experiences. The January 1967 event which initiated the letter to Allen Hynek, which in turn led to the complex MUFON investigation and Fowler's series of books, was the 'trigger event'.

Ray Fowler's second book in the series analysing the long and complex case of Betty Andreasson-Luca and her family was originally published in 1982, the same year as Budd Hopkins' *Missing Time*, but five years before Hopkins' landmark book, *Intruders*, with which it shares some common themes and methodology.

In many ways this second book about the Andreasson case is more interesting than the first, with its focus on one single extended abduction event from 1967. In *The Andreasson Affair, Phase Two*, Fowler and the MUFON team, including psychologist/hypnotherapist Fred Max, explore the lifelong abduction experiences of both Betty and her second husband, Bob Luca, and open out the discussion to what this phenomenon might be. Betty's determination to understand her experiences through the prism of her own Christian faith sets this case apart from many others and makes it problematic for some researchers; but Fowler, a Christian convert himself, nevertheless remains open-minded and keeps admirably to the scientific method despite the high strangeness of the data.

The detailed recall Betty has of her abductions under professional hypnosis continues to mark this case as significant—though the Hills' recall of their dual September 1961 abduction was also startling in its detail. A few outlandish and frankly horrific details—such as the abductors removing Andreasson's eyeball to insert implants into her brain, then re-seating the eyeball again into its socket—had never been reported to researchers prior to this case, but later came up again and again in other cases. (A notable example is the case of Ted Rice, whose long abduction history was explored by Dr Karla Turner.[8] When I corresponded with Ted over several months through 2012 he had never heard of Betty Andreasson, having never read any of Fowler's books.)

Betty's detailed artistic drawings of her abduction adventures are another aspect of this case which marks it out as unusual and interesting.

In the subsequent books of the series, Fowler tackles some of the more puzzling and enigmatic day-to-day phenomena experienced by the Lucas and frequently attendant to other abductees. *The Psychic Element in UFO Reports* stands on its own as a most fascinating essay: it examines the Pentagon's expressed interest in psychic phenomena experienced by members of a military helicopter team following an encounter, as well as poltergeist activity and temporary missing items (like keys) which inexplicably reappear in full view after hours or days, as is common to abductees.

Fowler explores theories about these and how they fit in with close encounters (CEs). Chapter Thirteen, 'Mystery Copters and the MIB', details Bob Luca's attempts to get answers from the FAA and the military about the black helicopters which frequently circled over their homes (in both Connecticut and Massachusetts) and also followed their car. The helicopters were unmarked, matte black, and flying well below the 500ft minimum altitude stipulated by aviation regulations. Many of Bob's close-up photographs of these black helicopters are reproduced in the book, as is his extensive correspondence with various authorities about the issue. He was certainly persistent, but without result. The weirdest phenomenon of all has to be the infamous Men in Black, reported consistently over years by witnesses, experiencers, and investigators and also a factor in this case.

In total, Fowler wrote and published five books on this landmark case: *The Andreasson Affair; The Andreasson Affair, Phase Two; The Watchers; The Watchers II;* and *The Andreasson Legacy.* The later books in the series go into the crossovers between abduction accounts and NDEs (Near-Death

Experiences) and relate 'messages to humanity' supposedly from the aliens, with a generally positive religious perspective.

Main Takeaways

1. Exhaustively investigated series of lifetime abductions of one individual, and later those of other family members

2. The first instance of the multi-generational nature of the abduction program, but because the investigators were focused on only the one case, no general conclusions about this were drawn

3. Radical and rarely detailed invasive medical procedures described, especially in the cranial siting of implants

4. Abductee a competent artist who illustrated her experiences in a series of detailed drawings

5. Abductee's religious beliefs heavily influenced her interpretation of the abduction events

6. The case spawned five consecutive books, all authored by Raymond Fowler

The Pascagoula Incident – October 1973

On the evening of 11th October 1973, work colleagues Charles Hickson (age forty-two) and Calvin Parker (age nineteen) reported that, while fishing off a pier on the west bank of the Pascagoula River in Mississippi, they heard a whirring/whizzing sound, saw two flashing blue lights, then observed an airborne, oval-shaped object 30–40 feet across and 8–10 feet high, descending towards them. Parker and Hickson claimed they were paralyzed while three creatures with "robot-like slit-mouths," and "crab-like pincers," floated down through the air and then took them aboard the object and subjected them to an examination. Two of the creatures escorted Hickson and one escorted Parker, effortlessly floating back up to their craft.

During the time they spent inside the object, neither Hickson nor Parker knew the whereabouts of the other. Parker claimed to recall nothing that occurred after the creatures were first sighted. Hickson says he later found Parker standing on the riverbank in a daze and shook him awake. There is some question as to whether Parker truly did remember nothing, choosing to feign ignorance, with Hickson's blessing, in the face of such a shocking and unbelievable incident.

Hickson and Parker initially attempted to report the encounter to nearby Keesler Air Force Base in Biloxi, only to be told: "The United States Air Force does not handle those things." They then went to the Jackson County Sheriff who, suspecting a prank, interviewed them separately, compared their testimonies and then (on the pretext of "going to get some coffee") left a hidden tape recorder running in the room where the two witnesses remained unaccompanied. The obviously very real terror expressed as the two talked of their encounter in private convinced the Sheriff that they were telling the truth about a real experience.

The pair described their abductors as around five feet tall, with elongated head but no neck, long arms ending in "mitten-like pincers," and legs which appeared locked together and never moved independently. Grey skin ran in horizontal folds around their bodies, and three pointed protrusions on their heads "could have been their noses and ears," whereas where the mouth was just a slit which never moved. No communication of any kind was reported with the beings by either witness, and both thought the entities might be robots rather than sentient beings because their movements seemed stiff and mechanical. The light inside the craft was reported by both witnesses to be so bright that the men were unable to focus on anything clearly.

Astronomer J. Allen Hynek flew down to Pascagoula to interview the pair and pronounced the case to be "absolutely genuine." The incident subsequently received widespread media coverage. Due to the unassailable character of the witnesses, their unshakeable testimonies, and the evidence which only became stronger the more deeply it was investigated, this came to be seen as one of the classic cases of multi-witness abduction.

Both Hickson and Parker voluntarily submitted to polygraph testing and to hypnotic regression sessions led by professional hypnotherapist John Kraus. One interesting detail is that both witnesses were very concerned they might have been subjected to harmful radiation (as had been Antônio Villas-Boas during his 1957 encounter). Due to the nationwide media interest and involvement of Dr Allen Hynek in the case, testing was carried out at Keesler

Air Force Base, where Hickson and Parker were given the all-clear. Finally interviewed by intelligence officers at Keesler, Hickson, and Parker reported that, "they believed our story, but Washington said 'Hands off'!"

The precise morphology of the entities reported by both Hickson and Parker has never been reported elsewhere; they did not correspond with the commonly-reported spindly grey alien type with bulbous heads and large, dark almond-shaped eyes. Hickson thought they might have been robots rather than living sentient entities, but he acknowledged that this opinion was speculative.

A book about the abduction titled, UFO *Contact at Pascogoula*, and authored by Charles Hickson and William Mendez was published by Wendelle C. Stevens in Tuscon AZ in 1983, ten years after the incident. It had only one print run and is consequently hard to find.

A footnote to the case is that many years later, several witnesses emerged who claimed to have seen the odd blue lights clearly as the craft descended towards the pier where Hickson and Parker were fishing.[9]

Main Takeaways

1. Dual abduction of two men outdoors, fishing off a pier on the Pascagoula River

2. Description of the entities seems unique, and the abductee witnesses thought they might be "robots"

3. The case attracted widespread national media coverage and was extensively investigated

4. As with the Villas-Boas case (the only point of similarity between these two cases), no missing time period was reported by the witnesses and Hickson recalled the events inside the craft in some detail, whereas Parker would not admit to remembering anything until he undertook regressive hypnosis

5. An early indication of how the federal government handled these cases: at Keesler AFB, Hickson and Parker were told "Washington said 'Hands off'!" so the investigation of the incident by Keesler was curtailed

Travis Walton – November 1975

Travis Walton was one of a team of seven forestry workers engaged in a tree-thinning contract near Snowflake AZ on 5th November 1975. At the end of their working day all seven men experienced a close encounter with a UFO while travelling home in a large pick-up truck on the mountain road.

On sighting the craft over nearby trees the driver, Mike Rogers, stopped the vehicle. While the other six loggers remained in the truck, Travis approached the craft on foot and was then seen to be struck by a beam of light and thrown some distance through the air. His companions, believing that he'd been killed by an "energy beam," fled in terror down the mountain road. However, within around 15 minutes, they'd regained composure and decided that they must return, if only to find Walton's body. But they could find no trace of him at the clearing where it had happened. They then went to the local sheriff to report the incident, explaining that Travis was now missing. The crew were met with incredulity and suspicion. Soon the police were quite openly suggesting that the truth was that one or more of the crew had murdered Travis, and that the rest of the crew had fabricated the UFO story to cover it up.

An exhaustive search effort through the mountains by police and local volunteers yielded no results. Five days after his disappearance, however, Travis unexpectedly turned up several miles away at the roadside, dehydrated and traumatised. He wanted to return to his quiet private life and adjust to his strange experience, but due to the excessive publicity surrounding his disappearance was reluctantly catapulted into the international media spotlight. Every living UFO investigator and group, including a number of professional debunkers, descended on Snowflake AZ, and as a consequence none of the witnesses had any kind of private life for years thereafter.

The case was extensively investigated by the authorities and by just about everybody else, and everyone has had an opinion about what really happened to Travis Walton. The controversy raged for decades, as the ideological bigotry against the whole subject shared by so many missionary-style debunkers refused to acknowledge either the proven facts or the credible evidence. All seven witnesses stuck to the story, incredible though it seemed, passed lie detector tests and went through a great deal of trouble as a consequence.

Walton wrote a book about the incident and all the events which surrounded it, initially titled, *The Walton Experience,* with later revisions titled, *Fire in the Sky.* I travelled to Snowflake AZ to meet Travis Walton in May 2010, and now own a signed copy of the 2010 edition of his book. A Hollywood film of the same title, directed by Robert Lieberman and starring D.B. Sweeney as Travis, was released in 1993 and became a modest box office success, despite a script which departed significantly from Walton's story.

Travis in person I found to be as intelligent, principled, straightforward and 100% genuine, just as most everyone else who has met him agrees. He told me that, had he known what trouble the abduction was going to create for him over the years, he would have stayed in the truck instead of getting out to walk underneath "the thing." He was quiet, reserved and suspicious of the motives of strangers who travel to Snowflake to visit him, as well he might be after all the harassment, bigotry and hostility from the likes of the late Philip Klass and his ilk, who worked diligently to trash his reputation. He knew virtually nothing of the details of other abduction cases in the public domain and demonstrated no interest in the general subject. It has been suggested that Travis might not in fact be a serial abductee in the program (see Chapter Three for details), but was abducted because the aliens had made some rash mistake in zapping him when he approached their vessel and then became determined to rectify their mistake and treat him onboard for the injuries they had caused. This is not entirely implausible, as Travis never prior to, nor following, the November 1975 incident has ever demonstrated any general interest in the UFO or abduction subjects and appears to carry none of the core psychological or experiential markers common to abductees as outlined in the long list in Chapter Three. ('Finding Out That You are One of Their "Chosen Ones"')

As of 2010 Travis continued to live a quiet family life in Snowflake. He had four children by his childhood sweetheart, whom he had married in the 1980s, and never discussed the 1975 incident with them. Moreover, he never spoke about it with anyone unless specifically asked and wishes it had never happened. I heard a few years ago that he had divorced and remarried but have no contact with him.

Abductions lasting for a duration of several days are rare, but not quite unique. Travis can recall only around 30 minutes of his five-day abduction with any clarity, so mystery surrounds the rest. It's possible he was unconscious for most of the time following injury suffered from the energy beam on the mountainside, but there is no way this can be known. A strong memory block may also be in place.

Main Takeaways

1. Abduction witnessed by six other people who had clear and close-up sight of both the UFO and the abduction (or at least of its initiation, before they scarpered)

2. Duration of the abduction was five days and nights, after which the abductee was returned famished and dehydrated but otherwise unharmed

3. Only 30 minutes of the five-day abduction was ever recalled by the abductee, so he was either suffering from a memory block or was unconscious during the remaining time

4. The case was extensively investigated and reported, and received international media coverage

5. The abductee had prior to his abduction demonstrated no interest in the UFO phenomenon, and post-abduction expressed no interest in examining or discussing other abduction cases

6. The event was made into a Hollywood film with several important details changed or amended to better serve 'the needs of the drama'

The Allagash Four – 26 August 1976

Four art school students devoted to the outdoors and with a shared attraction to wilderness adventure, identical twin brothers Jack and Jim Weiner with their friends Charlie Foltz and Chuck Rak, packed their station wagon with a Canadian-style canoe and camping equipment and, on 20th August 1976 left Boston MA for Northern Maine. The final leg of their journey required a small single-engine floatplane to airlift them, two at a time, into the remote wilderness of Telos Lake, astride the Allagash River. They planned to camp on the riverbank, canoe along the river, and do some fishing.

The first six days were idyllic if uneventful. On 26th August as they set up camp on the banks of Eagle Lake, that would soon change. They had decided to spend the early night hours fishing in the shallows across the lake from their campsite. They were several miles from the nearest human habitation and the evening was pitch-black, so the four built a large fire fuelled with enough logs to ensure a good three-to-four-hour blaze, to more easily find their way back to camp in the moonless night.

Not long after paddling the canoe out into the lake to begin their fishing, Chuck Rak began to get the strange feeling that someone was watching them. He would later state to MUFON investigators that as he turned around to scan his surroundings he saw a "large bright sphere of coloured light hovering motionless and soundless about 200 to 300 feet above the south-eastern rim of the cove."

He immediately drew the attention of the others to the spectacle, who each turned to see the glowing object coming up from behind the trees.

The object became larger and brighter. Each of the four would later recall how the object appeared to have a gyroscopic motion as though energy was flowing over its surface. The object seemed to divide laterally into four different colours which melted into each other "like a liquid."

Charlie Foltz picked up the flashlight from the floor of the canoe next to him and began to flash the light in the direction of the object sending an SOS, the only morse code signal he knew. The object's upward motion ceased and it began moving toward the canoe. A "tube-shaped beam of light" emerged from the craft reaching the water below and moved towards them; then the beam was on them and the object almost above them as they began frantically paddling to shore.

The next thing any of them remembered, Jack and Jim were back in the camp, Charlie Foltz was wading through the water from the canoe to the shoreline, and the strange object was moving away from them back across the lake. Chuck Rak at this point remained in the canoe, staring at the now receding, glowing sphere. Then Chuck appeared to come to his senses and left the canoe to join the others in the campsite. The large fire had burned down to embers, as though several hours had passed.

Despite—apparently—having been out in the canoe for just twenty minutes, and the anomaly of the abbreviated fire and the bizarre interruption to their fishing, they each turned in for the night, none of them discussing what they had just experienced. For twelve years following the incident the four had thought only that they had seen something strange and no more. They went their separate ways and got on with their lives.

Several years after the incident, Jim Weiner suffered a head injury and as a consequence developed tempero-limbic epilepsy. Persistent nightmares and brief snatches of terrifying memories began to emerge, and he eventually sought medical help. When he spoke to his brother about the nightmares he was astonished to learn that Jack, too, had been plagued by similar images. Eventually, Jim became convinced that the trip to the Allagash was the source of it all, and contacted his old friends, who agreed that there was something terribly odd about that long-ago incident on the lake. In 1988, he convinced them to join him in an attempt to figure out more, and the four were interviewed separately by both Ray Fowler, John Mack and a trained hypnotist. Dim memories were gradually recovered and the pieces assembled. Without the four ever discussing their separate interviews or hypnosis sessions, fantastic details began to emerge.

Each recalled being transported, floating, up the beam of light and into the craft, where they became immobilised but remained conscious. Their clothes were removed by the occupants of the craft, whom they each described similarly. Drawing from their art school training they each drew from their memories with some degree of professional competence: what they rendered were variations on the typical skinny grey with large head and very dark, almond-shaped eyes. They described having their clothes removed, followed by what seemed like a medical examination. Each had a sperm sample as well as other bodily fluids extracted. Chuck Rak described watching the procedures to which Charlie Foltz was subjected, though he alone does not remember undergoing them himself. Finally, their clothes were returned and they were ushered to a 'portal' through which they travelled back down the beam of light to the canoe. Full normality only returned when the four had paddled back to the shoreline where their camp was situated, with Chuck Rak being the last of the four to regain normal consciousness.

Despite the seemingly outlandish narrative, the separate memories were so strong and consistent as to challenge sceptical dismissal. Dr John Mack of Harvard University, who interviewed the four in depth, stated that in his opinion the case was a genuine instance of the rare phenomenon of the simultaneous abduction of multiple individuals. This is one of the strongest cases on record because of the exemplary character of all the witnesses, the consistently identical reports of the event from four separate individuals and the thoroughness of the investigation carried out over several years.

The fine illustrations of the event and of the entities encountered by the four are some of the best, mutually corroborative images ever of an abduction event. They are reproduced in full in Raymond Fowler's book, *The Allagash Abductions*, which also contains transcripts of interviews, photographs and maps of the area, and photos of the now well-recognised 'scoop-mark scars' on soft tissue areas of the body common to genuine abductees discovered on one of the witnesses. Indeed, both of the Weiner twins, and possibly Charlie Foltz as well, turned out to be serial abductees with a number of subsequent (and previous) abduction events in their lives.

Postscript

Chuck Rak claimed in a 2016 interview that the abduction "never happened," but did admit to the close-up UFO sighting and light beam over the lake. He claimed the other three had "made up the story" for financial gain,[10] despite the fact that no money was ever made, or even sought, from the story. Rak offers no explanation as to how his hypnotically recalled memories, and the drawings he made in 1988, matched so precisely those of the other three, as all had agreed not to compare notes.

Main Takeaways

1. Four men abducted simultaneously while fishing in a canoe, following a clear and close-up UFO sighting.

2. Two (twin brothers) and perhaps three were serial abductees; the fourth was probably not, but like Barney Hill, was opportunistically taken anyway though not subjected to the same procedures as the other three

3. As much as two or even three hours of missing time noted in hindsight but not examined for the next twelve years

4. Professional hypnosis to aid memory recall revealed identical narratives and descriptions of the abductors from all four witnesses

5. All four abductees were competent, professional artists and made detailed drawings of the abductors and the procedures to which they were subjected

6. Scoop mark biopsy scars and other tell-tale abduction markers were found on at least two of the abductees

The Debbie Jordan-Kauble Case – 1983, before and after

In September 1983, a young divorcée named Debbie Jordan wrote a long, detailed letter to researcher Budd Hopkins. The divorced mother of two small boys, who lived in a large family house with her parents in the suburbs of Indianapolis, had long been troubled by events of which she could make no sense. The letter initiated a series of investigations lasting more than two years and resulting in Hopkins' second book, a *New York Times* best-seller, *Intruders*. Although Debbie was given the pseudonym 'Kathy Davis' in the book, and the location of her suburban Indianapolis neighbourhood disguised as 'Copley Woods', her identity has now become known, as she later co-authored, with her elder sister (who really *is* called Kathy), their own book about this case.[11]

Together with Jordan's original letter were photographs of a large circular, burned patch of ground in the garden in Indianapolis where the soil had been baked hard, which had been discovered the morning following mysterious lights accompanied by an episode of missing time and her dog behaving in an uncharacteristically frightened manner. The large area of desiccated soil remained permanently barren of grass.

The investigation evolved into a complex case with some twenty individuals closely involved in, or witness to, a series of events spanning decades. Simultaneously, as a result of the public exposure he received following the publication of *Missing Time*, Hopkins had been approached by many other abductees from different parts of the country, whose experiences reinforced many of the new revelations being uncovered in the Copley Woods case. Two different NYC psychologists, other abductees working with Hopkins, and the film producer Tracy Tormé all became peripherally involved in the case.

This case is important because, for the first time, evidence of the alien breeding program and exactly how it operates was revealed and explored. Although many researchers had been uncovering parts of it, it was Hopkins who finally collated the evidence and began to make some sense of it. The case was also widely publicised and the strength of the evidence resisted all the predictable debunking attempts from the 'True Believers'. Debbie claims she had a pregnancy terminated during an abduction, and her offspring was brought to term and reared by the abductors.

Approximately five years after this traumatic incident during a further abduction, Debbie recalled an infant child being presented to her. The description of the "presentation" may be found in *Intruders*, including a drawing Debbie made of the child. She recounted this incident in great detail and alleges that she felt positive emotion from the aliens, tinged with some guilt, and that she felt the abductors "cared about her," and "were sorry." She was forbidden at this stage to have physical contact with the infant but stood several feet away from her and was told the child would not survive outside the carefully controlled environment the abductors had created. (i.e. aboard the UFO.) She later learned that the abductors claimed they had used her to breed several more children, and even told her she could name them.

In the course of his investigations, several male abductees recounted to Hopkins the manner in which sperm is regularly harvested from them, and how important this process seems to be to the abductors. Some new information about the nasal implants placed by the abductors was also uncovered, as are many of the more mysterious and hitherto little understood aspects of the phenomenon. The revelation of all the puzzling evidence and the conclusions drawn had profound effects on the subsequent direction of the entire field of study and brought many new researchers into the field.

Intruders was scripted as a TV mini-series by Tracy Tormé and eventually made into a feature film of the same title, directed by Dan Curtis and starring Richard Crenna, who plays an amalgam character of Hopkins and John Mack. The cast also included Daphne Ashbrook, Steven Berkoff, and Mare Winningham. The *Intruders* mini-series/film is not a literal transcript of the book's narrative but a re-written screenplay exploring the main themes and with some changes to the characters—who nevertheless remain largely recognisable—to better fit the mercenary requirements of the film industry. During the personal time that I spent with Hopkins, he told me he "gave it a B+" for its accuracy in dramatizing the phenomenon. (He thought parts of the story were a little over-sensationalised and one or two important things were missed out, but it was OK overall and worked well as a drama.) The film was moderately successful, introducing millions of people to the abduction phenomenon who for the first time began to see it as a credible reality.

Main Takeaways

1. Continued abductions through several generations of the same extended family over decades

2. A rare example of a UFO landing to initiate an abduction, with a significant and scientifically measurable impact on the local

environment and vivid recall from the abductee, who was able to draw the unusual, egg-shaped UFO

3. Medical evidence of the placing of nasal implants in abducted children

4. The precise methodology of the gestation and rearing of hybridised alien-human children was revealed, where the human mother is later abducted to 'bond' with the child following the removal of the foetus from her during an earlier abduction. Hopkins cites several different cases of this phenomenon, all of them confirmed by hospital obstetricians or gynaecologists

5. Very detailed descriptions of how sperm samples are acquired from male abductees, and the high importance this procedure seems to have to the abductors

6. Close-up, fully conscious interactions/sightings of small greys in an abductee's home with the abductee fully conscious and able to move: this is so rare as to be almost (but not quite) unique

7. Repeated interactions with what seem to be adult hybrids in the 'everyday' world, including one episode with a six-hour period of missing time and at least two with multiple human witnesses, each lasting many hours

Each of the selected abduction cases detailed above found their way into the public sphere, and often into national or international news reports. Thousands of other cases, some not reported until years after the event, occurred simultaneously with them. Patterns began to emerge: the cases had far more similar, often identical features than notable differences.

The next chapter looks at what all these abduction reports were telling us. In sharp contrast to the idea that extraterrrestrial visitors were 'investigating', 'examining', or 'studying' humans to find out about their anatomy and physiology, it began to look like there might be *a program* involved.

Chapter Three

The Program

What is it, and why describe it as a 'Program'?

When researchers began to examine the abduction phenomenon in the 1960s, the similarities between individual cases were striking, whereas the differences between them were minor. After the meticulous analysis of many thousands of events through the past fifty years, it is now clear what a 'normal' abduction includes and, just as importantly, what it does not.

The abductors' motivation and purpose for initiating each event does vary somewhat depending on the abductee, often leading to what are sometimes referred to as secondary or ancillary procedures, but certain primary procedures are almost always carried out in the same sequence on all abductees, on virtually every occasion.

Historical Overview: What People Report Happens to them

In the late 1970s, Budd Hopkins investigated two abduction reports dating from the 1920s and several others from the 1930s. Since the 1960s when the Villas-Boas and Hill cases came to public attention, and occasionally prior to that, tens of thousands of people from all walks of life and from every location on Earth have reported extraordinary but near-identical personal experiences. Some people have provided anecdotal evidence, from family history, which suggests that these events had been occurring with previous generations back to the 1890s or even earlier, so traumatic that the witnesses retained the memory of a missing time period and the 'I know something strange happened that day' feelings which could never be wished away nor forgotten. These under-reported cases lurk beneath the surface in a large population of abductees, but are generally never publicly discussed because, what do you say? And to whom? How would you even begin this conversation?

These people report being taken suddenly, with little or no warning, out of their ordinary lives by strange beings; transported to an obviously artificially constructed interior space and placed on a plain, hard horizontal surface which reminds them of a medical examination or operating table, then subjected to a range of physical and mental, quasi-medical and often anatomically invasive procedures. The person is then returned to his/her precise location immediately prior to the initiation of their abduction, where almost all of them promptly forget everything that happened and retain only bewilderment at having two hours or more of missing time, of which they can recall nothing except a nagging and uncomfortable feeling that 'something weird happened'. Imagine trying to explain that in 1929!

Following the experience, marks and scars are often discovered on the body: Straight line scars, long and prominent; scoop marks on soft tissue areas like the back of the thighs or calves; precise triangular patterns of fresh puncture marks and numerous other patterns; unusual bruises and burns. None of these have a prosaic explanation and invariably result in puzzlement but are usually shrugged off with thoughts like: "I can't remember how I did that," or "I just woke up with it: I must have bashed it on the bedpost while sleeping," or other similar reasoning which we use to persuade ourselves that "everything is normal" when we're in fact confronted with the extraordinary and unexplainable.

The abductors have for decades consistently been reported to be thin, spindly humanoids, hairless with abnormally large heads for their bodies and enormous, almond-shaped, very dark eyes with no detectable pupil or white area. They are described as having a small slit for a mouth, a tiny vestigial nose and a hole on each side of the head where human ears would be. They are usually reported to appear as one of two types: short and tall, but otherwise nearly identical.

The skin colour is usually described as light grey but occasionally as dark grey, tan, and very pale white. In the famous Kelly Cahill abduction near Melbourne Australia in 1993, the creatures were reported to be black, though otherwise with identical body morphology to the more common greys. The entities witnessed by the children at the Ariel School incident in Zimbabwe in 1994 were similarly described to Dr John Mack as spindly entities with large heads for their bodies, big almond-shaped eyes, and completely black bodies including their heads and hands. (Both Australia and Zimbabwe are in the

Southern Hemisphere. This might be just coincidence and we shall not speculate further, merely remark on the fact).

The shorter ones—generally described as between 3ft–3ft,6in (~1 metre) in height—usually work in groups of three. They retrieve the abductee from her/his normal environment then transport him/her, paralysed and helpless, to the examination site by some kind of levitation. Many of these people report being "pulled up a beam of blue light" to a large stationary object (a UFO), which they are able to clearly describe as they approach it, and often recall seeing the rooftops of their neighbourhood buildings below as they ascend.

On arrival, the beings sometimes assist the abductee in removing his/her clothing—including all jewellery and wristwatches or anything else worn—in a 'waiting area'. The person is then escorted naked to what seems to be an examination room. Some abductees report being taken directly to this room fully clothed and undressed when they arrive there: their clothing is then left on the floor while the procedures are carried out. Doorways are described as smooth, straight-sided apertures curving at the top and sufficiently tall for the average human to walk through upright.

The examination room is fitted-out with one or more rectangular tables with smooth, rounded corners, large enough to accommodate the tallest human. Sometimes the room contains a single table, more often a few more, and occasionally in some rare cases, a great many of these tables are reported inside a very large interior space with a smooth, curved, high ceiling. No corners are ever described but rather the walls, floor, and ceiling smoothly blend into each other without edges. The space contains no colour or decoration, no art or artistic design features: everything is described as plain and neutral—strictly utilitarian. The predominant tones of everything are white and grey. Light diffuses "from somewhere" to illuminate everything but no light fittings are visible: the room lighting is "just there; it kind of comes from the walls," is how abductees often describe this effect.

The Medical Examination

There are instances when a medical examination is not conducted during an abduction, but they are outliers. These rare cases are usually instances where a very recent abduction and examination have taken place (like the day before, for example) and this second abduction is of short duration and appears to be for some specific follow-up purpose.

Two or three, or occasionally four, of the small beings together carry out a physical examination, which may take fifteen minutes or so. The beings are focused on the task, at which they work silently and efficiently. They typically start at the feet and work upwards, but some abductees report the beings starting with the head and working down the body, then working back up again.[1]

They poke and palpate the soles of the feet, check the reflexes by drawing a metal instrument over each sole, rotate the ankles, flex the knee joints, squeeze the calves and occasionally make an incision on the calf, thigh or arm which can be either a small hemispherical scoop or a thin straight cut, usually an inch or two (up to 5cm) but occasionally reported as longer, even up to 12 inches (31cm) in length which often forms lifelong scar tissue. Female abductees report a gynaecological examination; sometimes the uterine wall is scraped with a long instrument and a sample collected, followed by a breast examination. With male abductees, the genitals are always palpated and inspected.

The head examination is a major focus. The head is held between both hands of one of the small beings and moved around (possibly to check the flexibility of the cervical vertebrae); ears are examined with a long instrument and eyes inspected with a light-emitting device. They inspect the throat, teeth and gums, may take samples of gum tissue and always insert their thin bony fingers into the mouth. Sometimes a wad of material is pushed into the mouth, making the abductee gag and choke before it is withdrawn and removed from the room, presumably for analysis. They seem to pay close attention to the thyroid glands and the neck.

The abductee is then rolled onto his/her left side and the spine is carefully checked from the cervical vertebrae to the coccyx, one vertebra at a time, by one of the creatures using his fingers—a process which may be repeated several times by different members of 'the crew' as if to ensure nothing is missed. Abductees report that this spinal examination component is most careful and thorough and takes the longest time of any of the procedures.

The subject is then rotated onto his/her front and a rectal examination is carried out, usually with an instrument shaped like a small wire whisk which is inserted then withdrawn, but occasionally with larger instruments or devices. The small beings check meticulously any new marks on the skin for signs of recent visible injuries: abductees report that such things as bruises and scabs formed following minor accidents, minor fractures incurred from a fall, teeth recently extracted or even dental braces being fitted since the last abduction, always attract the small beings' attention. (As mentioned above, Barney Hill reported that the beings were much concerned as to why he was fitted with removable false teeth, and why his wife Betty was not.)

Often at the end of the examination the beings implant a tiny, round metallic object in the ear, nose or sinus cavity, or else they remove one placed during a previous abduction. These devices will be discussed in more detail in Chapter Four. The abductee is usually, but not invariably, then told to lie on his/her back on the table, and one of the taller beings enters to assume control of the more complex procedures. (More on these later.) Sometimes retaining straps are used over the upper arms and lower legs, which are described either as "metallic," or "like tough velcro," but the presence of these restraints is not always described. Some of the examination tables are reported to tilt through 90 degrees to allow for the abductee to be 'scanned' with a light-emitting device, so restraints may be required for this purpose. Also, the record of compliance or physical strength of the abductee, or his/her occasional ability to break the partial paralysis and neurological control, may in some cases be a factor. I recall one incident during which I felt strong fear and anxiety from one of the small greys when dealing with me, so there may be a history of defiance or anger. I have photographs of bruises and red chafe marks discovered around my lower legs on waking from sleep. (See Appendix B.)

Types of Beings Most Often Reported; Telepathy

The taller beings are relatively taller than the shorter ones and described as standing between 5ft–5ft,6in (~1.6 metres), but never more than that. The taller type seems to be in charge of the shorter ones and, curiously, is often reported as being a little more forthcoming when questioned, though usually only responding with bland reassurances like, "You'll be OK" or, "We're not going to hurt you" or, "This won't take long".

Interestingly, the taller beings are usually described as being either 'male' or 'female', although no sexual organs or secondary sexual characteristics are ever seen or described. The abductees merely report that some of the taller ones seem male and some seem female, and they can tell the gender difference easily when telepathically addressed by the beings. Curiously, all the smaller grey beings are usually described as being 'male' in nature: many abductees are definite about this and are certain the gender issue is unambiguous: the small greys are all 'guys' and none of them are 'girls'. A small number of abductees however think that the small greys seem neuter or sexless, or even that they are robots or automatons of some kind.[2]

All communication—between the beings themselves, and between them and abductees—is telepathic. Occasionally another abductee will cry out audibly during a procedure or something will fall to the floor with a crash, and then the beings visibly react to the sound, so can obviously hear sound waves as we do. The creatures never audibly 'speak' and presumably never need to, or perhaps lack vocal cords so are unable to. Abductees report that they 'hear' in their heads when one of the beings addresses them and always understand, and when they respond mentally they can 'hear' their own response and know they are being understood.

This is precisely my experience: each being has a distinctive 'voice' and you can tell them apart when more than one is engaged with you during a procedure and one responds to a question, because they very rarely communicate voluntarily except in anodyne palliatives such as, "You'll be OK" or, "We won't hurt you".

Abductees frequently report that they understand the beings when they 'talk' between each other when in close proximity to the abductee, so the telepathic communication does appear to be constrained or weakened by distance, like our sound-based speech. Although all the beings have small mouths visible, there are no reports of the mouths ever seen to be used or even opened. When one or more of the beings are up close to the abductee, there is never any evidence of breath being inhaled or expelled, and the very slender, straight chest area is never seen to move, so they probably don't have lungs or breathe air. From their body morphology, it doesn't seem that these beings ingest food via the mouth or have an alimentary canal or digestive system.

A German abductee I know told me the abductors always communicate telepathically in the German language to her, so it's possible that one of the advantages of direct mind-to-mind telepathic communication is that the need to learn languages at all is redundant and the understandings somehow flow directly from brain to brain.

The abductors are always reported as task-oriented or business-like, rarely betraying any trace of emotion, though the taller beings are occasionally described as exhibiting mild irritation with their smaller co-workers if some menial task is not performed quickly enough. They can express mild satisfaction or gratitude when the abductee is compliant and helpful, or annoyance and irritation, but never lose their temper or become angry. Their behaviour speaks of a familiarity with their tasks and everything seems routine, like working on a production line. Though they communicate telepathically with each other and with the abductee they are not usually voluntarily communicative about what they are doing or why, although occasional exceptions to this are not unheard of.

Abductees report they are certain the same beings working in a 'team' deal with them again and again, over years of abductions, so in some cases they feel they 'get to know them'. However, although the beings are sometimes insistently inquisitive about what the abductee is doing in their normal lives, especially concerning issues which affect the abductee's health in one way or another (they don't like it if the abductee is a smoker, for instance), the abductors themselves seem to have no life outside their service to the abduction program—i.e. they don't appear to play the piano, go out to restaurants, play recreational sports or have any identifiable family life. They just seem to work, incessantly, on an endless series of abductees, and do nothing else: it seems to abductees that they are designed and programmed for this single activity. We shall explore what these beings really are, and their likely origins, in the concluding Chapter Nine.

Beings who appear human-looking, or even who are nearly indistinguishable from normal humans, have also been reported, increasingly so since the 1980s. Also occasionally reported are taller insect-like beings, usually described as standing around seven feet (more than two metres) tall and "looks like a big praying mantis."

These, of course, don't look remotely human but are invariably described as having the strongest mental control at the greatest distance over the mind of the abductee and who are deferred to by all the other types of beings. Perhaps these may be the actual instigators of the program, for whom all the others work. We'll discuss these different types of beings in more detail later.

Control Over Human Neurological Systems

For the entire duration of the abduction, subjects claim to have no voluntary control over most of their movements: they are unable to attack, kick, or punch their captors. In very rare, exceptional instances the 'mind control' seems to be temporarily overcome and the abductee attempts to flee before recapture and control is re-established. Most of the abductees' neurological systems seem to be controlled very effectively by the abductors, preventing movement excepting those which the captors need the abductee to perform, such as sitting or lying down, standing up and walking slowly while escorted, or moving their eyeballs. Although in many cases the eyes are unaffected and may move freely and voluntarily throughout most of the abduction, there is a notable exception to this general rule which we'll investigate later.

The degree of control over the human brain, neurological pathways and motor functions is very extensive and ensures a high degree of involuntary compliance from abductees who have no choice but to submit to the abductors' commands and intrusive procedures, irrespective of their remonstrations or pleading, which are ignored. When asked questions like, "Why are you doing this?" the (telepathic) response is either one of the usual anodyne palliatives such as, "You'll be OK," or else some variation on, "Because we must" or, "We have the right," or even drily with deadpan sarcasm, "You are one of our chosen ones."

The Persistent Strangeness of it All

The events reported by these witnesses, all over the world, are utterly implausible and, superficially, physically impossible. People report being immobilised, while fully conscious, and transported through closed windows and doors, car windshields, walls, and even the ceilings and floors of city apartment and office blocks. Although many such events take place in towns and cities, where the hovering craft ought to be plainly visible to thousands of onlookers, they are very rarely seen by others—not actually never seen, but very rarely. Depending on the time of day and location, the abductee is sometimes missed and, in any case, can never be located during the period of their abduction.

These thousands of abductees worldwide are from diverse social, political, cultural, religious, educational, intellectual, economic, ethnic, and geographic backgrounds. They are doctors and every type of medical and paramedical professional, psychologists, businesspeople, educators, artists and musicians, scientists, social workers, childcare professionals, truck drivers, shop workers, laboratory workers, office workers, policemen and women, serving members of the armed forces, farm workers, factory workers, labourers, full-time homemakers, university students, and the unemployed: a complete cross-section of society. What they all describe has no antecedent in popular culture. The reach of the phenomenon is global: people describe the same things in the same detail worldwide. This phenomenon is reported with perplexing consistency, forensically identical in detail, and is very widespread.

As so many researchers have pointed out, if this is not happening in reality, but is the result of some mass psychosis afflicting the entire world, then that warrants a serious investigation by the global professional psychological and psychiatric community, as it would be without question the most significant and long-enduring phenomenon ever known in the history of the social sciences.

Memory Blocks and Missing Time

One of the abductors sometimes tells the abductee that they are not going to remember anything about what happened. (As mentioned above, I personally have been told exactly this in a bored, matter-of-fact way.) On return to their normal environment, the abductee notices that one, two, or even more hours have passed of which they have no memory, and the 'missing time' is seamless.

For example, a woman might be walking on a woodland track and suddenly find herself standing still, a hundred metres further on in the direction in which she was walking. This has often been described as like an old-fashioned celluloid film which is cut and re-joined: one second you're in one place and a split second later you're on the other side of the field, or several miles further on the road. She discovers that two hours have passed and cannot account for the missing time but has an uneasy feeling that "something strange has just happened."

Or a couple driving in a car see a large bright light in the sky and pull over to the side of the road to "take a better look at it." What seems like a second later, they find the light is no longer visible and they discover that more than two hours have passed: they both have an uneasy feeling that "something happened that night which we can't explain or remember." Occasionally they find themselves again driving in their moving vehicle when this realisation dawns on them, but neither can remember actually starting out again from where they stopped to watch the thing in the sky.

I myself had a similar experience in the spring of 1997 while driving alone from Reading in Berkshire, where I had attended a business meeting during the afternoon, to Amersham in Buckinghamshire, where I had an early evening appointment. I took a road that I didn't know well but parts of which I had travelled on maybe two or three times previously. Leaving Reading around 5pm, in full daylight, I headed off towards Amersham, which I had expected to reach within an hour. The journey was odd in that the road was deserted, which is a rarity in the Thames Valley, especially at that time of day. The only vehicle I saw was what appeared to be a large, black or dark-coloured truck or bus coming towards me, seemingly in the middle of the road, with its full-beam headlights switched on and moving very slowly. I slowed to avoid a collision and, the next thing I knew, I was in a traffic jam going into Amersham in the dark. The time on the clock showed 7:35, and I had missed my appointment. What the heck happened there?

I had an hour's drive back to Ewell in Surrey where I lived at the time, and on undressing to take a shower while still thinking about the strange journey, discovered that the buttons on my shirt were misaligned. Moreover, a treasured piece of jewellery—the only one I ever wore—was missing. I have not since found it. It could not have been accidentally left anywhere because it was never removed during the day: a silver artifact on a strong silver chain around my neck and worn under my shirt, it was only removed at night and was never on open view.

At this time in my life I still did not suspect that I might be an abductee. Besides, my late mother's repeated attempts to engage me with esoteric subjects had been met with resistance and incredulity on my part. The idea that something supernatural had been interfering with both our lives was all too weird for me. At the time, I had no idea how to process that even if I could be persuaded that it was true. My mother when alive didn't consider the possibility that she and I were 'abducted by aliens', or at least never said as much to me, but like her mother before her, knew something abnormal ("the pixies" was how my grandmother interpreted it) was interfering with her life.

It's now obvious to me that my silver chain and amulet were removed along with everything else during the abduction in preparation for the usual examination, and not returned when I was re-dressed prior to being put back in the car. Presumably another abductee found he/she was wearing it and is still puzzling over where it came from on that day. Occasionally minor mistakes are made by the 'small greys'; items of jewellery and clothing are mixed up and returned to the wrong abductee, and this seems to be an example of precisely this happenstance.

It is most common for people in these circumstances to have no memory at all of what happened during the event. Far less common is for some fragmentary memories of the first or last few seconds of the experience. A very small minority of abductees claim to recall all, or most, of the abduction episode. It is obvious that some process of enforced amnesia is at work, but the process is not infallible and some memories occasionally bleed through. Almost everyone—certainly those experiencing a daytime event—notices the missing time even if nothing else is recalled, and has an uneasy feeling that 'something strange happened'. Night-time abductions, however, might easily go almost completely unnoticed other than the awareness of an unusually vivid dream.

Other Essential Facts About the Phenomenon

1. Abductions are not random events but seem planned as part of a program. Abductees report that occasionally the taller aliens—and more especially the rarely seen mantis-like beings—reveal this to them in their typically offhand, matter-of-fact manner. The widespread reach and scale of the phenomenon displays purpose, planning, and the investment of resources beyond anything which might be thought of as casual or random. It renders utterly implausible any notion that the beings are 'merely curious'—that they are conducting some kind of research into the human race. They know exactly what they are doing. Their understanding about, and almost total control over, human physiology and neurology is incomparably greater than our own present knowledge, and every single abduction event is obviously goal-directed and purposeful

2. The phenomenon is intergenerational: the children of abductees often report being abducted themselves, as do their children, and so forth. This kind of genetic continuity seems to be important to the aliens, though we have little idea why this should be. We'll explore some ideas on this in Chapter Eight

3. Abductions are initiated in childhood and repeated at regular, and often frequent, intervals until old age. There is some evidence that when abductees reach a very advanced age, the abductions become rare events or even stop altogether, perhaps because the elderly are no longer of any use to the program

4. Occasionally, subjects report a series of abductions with short intervals between, like one (or even more) each day for several weeks. At other times, the same abductee will report the interval between incidents to be much longer, like only one single incident every several weeks, or even months

5. Abductions occur at all times of the day and night, depending on access to the abductee and when they will least be missed. They may occur 24 hours a day, 7 days a week

6. Abduction activity is unrelated to the consumption of alcohol or drugs and has no correlation with psychosis or mental illness

Some Thoughts From Budd Hopkins

This remarkable man will be discussed in more detail in Chapter Six, as during the final two years and eight months of his life I had the good fortune to get to know him well and was privileged to spend a great deal of time alone with him. However, here we shall introduce some poignant thoughts of his regarding this phenomenon and how different people react to it. He was highly intelligent and unusually perceptive, as the two passages below reveal.

In a 'Note to the Reader,' at the beginning of his seminal 1987 book, *Intruders*, Hopkins introduces the subject as follows:

> Whether you are a physicist, a housewife, a UFO researcher or a dabbler in the occult, this book will almost certainly strain your credulity to the breaking point. One of the many things we don't like to admit about the human mind is its basic inability to accept or even to vividly imagine an 'unrealistic' or deeply unpalatable truth. Though we can entertain almost any wild idea 'in theory', a profoundly unsettling concept can be impossible to believe—to really believe—despite the weight of evidence and pressure of logic.
>
> One historic example of our inability to comprehend and believe a chilling truth is delineated in Walter Laqueur's book *The Terrible Secret*, a work dealing with the Holocaust. Laqueur's research established that by the end of 1943, when a sizeable portion of the world's population had read or been told of Hitler's systematic liquidation of the Jewish people, this ongoing horror was simply not believed. The Nazis were evil, we seemed to be saying, and truly barbaric, but *that* ... THAT just could not be true. Obviously, in this context even eyewitness accounts were irrelevant. Laqueur describes a meeting between Jan Karski, a Polish eyewitness to the slaughter, and Judge Felix Frankfurter, a man whose brilliance and intellectual resiliency cannot be doubted. Karski told Frankfurter what he had seen and heard [about the systematic, organized and widespread mass

murder of Jews in the Nazi-occupied territories of Eastern Europe], but Frankfurter replied that "he did not believe him. When Karski protested, Frankfurter explained that he did not imply that Karski had in any way not told the truth, he simply meant that he **could not believe him**—there was a difference." [3]

Hopkins continues:

> Obviously I do not mean to compare the unspeakable horrors of the Holocaust with the events reported here. An analogy exists only in the methods we use to avoid such deeply disturbing testimony.
>
> A majority of the world's scientists believe in the probability of extraterrestrial life existing somewhere in our inconceivably vast universe, and that some of these life forms are possibly more advanced than our own. Many scientists in fact retain an active interest in SETI ... and yet almost none have taken the time to look into the UFO phenomenon as it unarguably exists: consisting of tens of thousands of reports of apparent craft sightings, landings, photo and radar evidence and accounts of the temporary abduction of human beings.
>
> There is an all-too-human reason for this lack of curiosity. The idea of an extraterrestrial intelligence existing 'out there' somewhere ... is an easy, logical and comforting concept to hold. We on Earth remain, according to this model, detached and unaffected, passively listening for distant, intelligent signals sent to us across the 'unbridgeable vastness' of space. The possibility that extraterrestrial intelligence may already be visiting our planet, as the UFO evidence implies, and treating the human species as laboratory specimens for some elusive and unfathomable purpose—that is a truly disturbing idea. We all know, of course, a basic scientific truth: it can't happen here. Justice Frankfurter's remark is apt: despite the eyewitness descriptions and all the other categories of evidence, **"I just cannot believe it."** [4]

Hopkins had over the years become used to dealing with those who were 'skeptical' about the UFO phenomenon. This, in itself, wasn't particularly surprising. What annoyed him were the high profile public figures who were often given a TV platform "to present the skeptical point of view" such as Seth Shostak, James Oberg, the late Philip Klass, and others of their ilk who refused to acknowledge or give credence to eyewitness accounts—often mutually corroborated—or photographic, cinematic, and radar trace evidence of UFOs presented by military and civilian pilots or other sober professionals. No evidence, no matter how compelling, would ever shift these committed ideologues, whose mindset went well beyond mere skepticism to embrace a never-ending parade of so-called 'rational' explanations which were so ridiculous as to beggar the question of who the real lunatics were.

Sitting at his kitchen table one morning in October of 2010, enjoying coffee and almond croissants from West Side Market, Budd told me that, in his long experience, people can usually be seen to adopt one of four attitudes when confronted with the UFO and abduction experience:

1. **The genuine skeptics** are prepared to examine the evidence without prejudice and may be characterised as having an essentially scientific approach to the issue. Their thought processes are logical and they are driven by the evidence, however uncomfortable, frustrating, or unexpected it may prove to be. They are not fooled by hoaxers and fraudsters and are able to 'separate the signal from the noise'.

2. **New Age Positives** are by nature optimists and fully acknowledge the phenomenon to be real, gushing about the benign 'space brothers and sisters' visiting us from the stars. To them, these are more spiritually evolved beings come to help us advance to a higher state of consciousness before we destroy ourselves. Often, this scenario includes some form of confrontation with sinister cabals of human 'elites'—the Shadow Government, the Military-Industrial Complex, the Trilateral Commission, or some such trope—who resent their impending loss of power and control and are determined to thwart the good intentions of our Galactic Saviours. The New Age spin on the abduction phenomenon is always positive: the aliens are concerned with the survival of Planet Earth. They are committed environmentalists, like little cosmic Greta Thunbergs, and are in the process of 'upgrading' humanity to its next evolutionary level. They are never, ever the Bad Guys: that role is always strictly reserved for the sinister humans attempting to protect their miserly and destructive status quo.

3. **The True Believers** just know that there are no genuinely unidentified flying objects, as such things can always be explained

away as swamp gas; optical mirages; meteors; astronomical phenomena; lighthouses; the planet Venus; misidentified aircraft; and, of course, hoaxes. The abduction phenomenon obviously doesn't exist because it cannot, given that our planet is not being visited. It has to be sleep paralysis (even when the abducted are awake and driving in company with other passengers on the highway); hypnagogic or hypnopompic states; suppressed memories of sexual abuse as children; psychotic delusions. And, of course, hoaxes. Physiological evidence such as scars, scoop marks, and retrieved implants from the bodies of abductees are ignored and brushed aside. Hopkins compared this mind-set to that of religious fundamentalists, observing that "the True Believers believe that UFO abductions don't exist with the same fervor and religious certainty with which the Pope believes in the Virgin Birth."

4. **The Incurious**, when confronted with evidence for the extraordinary, shrug their shoulders and move on. They can generate no interest in the subject, and see no implications for their own lives, so can't be bothered to engage with it.

Had I not been inducted into the program in infancy, but instead been free to live a 'normal' life, it's a mildly diverting point of speculation into which of Hopkins' categories I myself may have fallen. People who know me well seem to think that I would be one of the incurious. My natural interests lie in art, music, the study of history, aviation, and travel. It is chiefly those subjects with which my mind and spirit would live their lives and be happy. I never wanted to be part of this. It took me years to accept it, but I wish they would abandon their program, pack-up and go home (wherever that is) and leave us to work out our own problems by ourselves. We don't need no stinkin' abduction program.[5]

... and Introducing John Mack

John Edward Mack, MD (October 4, 1929—Sep 27, 2004) was an esteemed professor of psychiatry at Harvard University Medical School whose natural scientific curiosity drew him to study the alien abduction phenomenon during the last twenty years of his life. Indeed, he lectured widely on the subject. After meeting and working with more than 200 abductees whom he had encountered as patients in his psychiatric practice at Cambridge Hospital in Massachusetts, Mack wrote and published two books on the phenomenon: *Abduction*, and *Passport to the Cosmos*. He was also the recipient of the Pulitzer Prize for his seminal biography of T.E. Lawrence, *A Prince of Our Disorder*.

In his foreword to Dr David Jacobs' 1993 book, *Secret Life*, Dr Mack writes:

> The idea that men, women and children can be taken against their wills from their homes, cars and schoolyards by strange humanoid beings, lifted onto spacecraft, and subjected to intrusive and threatening procedures is so terrifying and yet so shattering to our notions of what is possible in our universe, that the actuality of the phenomenon has been largely rejected out of hand or been bizarrely distorted in most media accounts. This is altogether understandable, given the disturbing nature of UFO abductions and our prevailing notions of reality. The fact remains however, that for thirty years and possibly longer, thousands of individuals who appear to be sincere and of sound mind and who are seeking no personal benefit from their stories have been providing to those who will listen consistent reports of precisely such events. Population surveys suggest that hundreds of thousands and possibly more than a million persons in the United States alone may be abductees or "experiencers", as they are sometimes called. The abduction phenomenon is, therefore, of great clinical importance if for no other reason than the fact that abductees are often deeply traumatised by their experiences. At the same time the subject is of obvious scientific interest, however much it may challenge our notions of reality and truth.[6]

Dr Mack visited Ruwa in Zimbabwe following the 1994 'Ariel School Incident', and interviewed many of the child witnesses to the event there on camera (Mack specialised in child psychiatry at the Cambridge Hospital, MA). He and his work is one of the main themes in Randall Nickerson's 2022 documentary film, *Ariel Phenomenon*.[7]

As Hopkins and Mack explain so lucidly in the forgoing passages, the thing which stops most people from taking the abduction phenomenon seriously is that, against our current preoccupations and accepted notions of consensus reality, it just seems so *unlikely*, so *improbable*: "I simply cannot believe it, regardless of the evidence," sums it up. Serious engagement with the subject is, furthermore, ill-served by its trivialisation in popular culture, which has the effect of consigning the subject to the fringes of populist entertainment, far from the centre of gravity of serious scientific enquiry.

The evidence reveals that such attitudes serve the aliens' agenda very well. The secrecy of the program is paramount: it is obviously of vital importance to these beings that our civilisation continues to ignore what they are doing and attempts no collective action to stop it. The secrecy extends to:

1. The ability to somehow prevent them or their craft being seen as they abduct people
2. The ability to prevent the abductee being seen as s/he is moved
3. Evidence of profoundly advanced neural manipulation that allows the abductors to control the actions and perceptions, and block the memories, of their subjects

Finding Out That You are One of Their 'Chosen Ones'

Abductees report a range of very specific experiences during their lives; many of the most common are listed below.

Some of these are persistent, even lifelong feelings or convictions, which cause the abductee to have certain attitudes or behave in a certain way: like a conviction that they are somehow 'different' or 'special'.

Some are sudden and memorable singular events: like a compulsion to walk or drive to a specific location at a specific time and on a certain day, for no discernible reason.

Others are 'weird stuff which happens around me', such as seeing balls of light in the house, or affecting electronic gadgets and streetlights when walking by.

This list may not be comprehensive but to even an 'unaware abductee' many of these will likely set off bells of recognition.

1. During their lives they often admit to having experienced periods of lost time, from an hour to several days, during which they cannot recall what happened to them or what they were doing

2. Many report fear or unease when they travel down a particular stretch of road, for no obvious reason save for a vague impression that

something strange and disturbing had happened to them there in the past. For some this phobia is so severe that they will often take a detour of several miles

3. A compulsion to fulfil a secret 'mission' in life but don't know exactly what it is; a deep feeling of having a 'calling' that one day will be revealed, or a feeling that they are somehow 'special'

4. As an extension of point 3, some have a deep feeling that one day a great cataclysm or 'change' will happen in which they will be required to act or assist, and when that day arrives they "will know what to do"

5. Chronically low self-esteem is very common. They often lack confidence in their abilities and find it difficult to engage successfully with a career, despite being assessed as a child to have high intelligence ($IQ \geq 110$) and delivering an 'above average' academic performance at school

6. They develop psychic abilities or an interest in them, and feel ashamed or embarrassed about admitting this

7. Many become attracted to religions, mysticism, spiritualism or the occult and sometimes investigate these areas of belief to the degree that they join the local congregation, temple, or group. After a time they leave and move on, because they find no answers to the core of what is troubling them

8. Attraction to or even commitment to environmentalism is common: some become politically active and many become vegetarian

9. Many admit to having powerful recurrent dreams of planetary thermonuclear war or large-scale destruction from environmental disasters

10. Many report recurrent dreams of flying through the air or levitating, emphasising that these 'dreams' are accompanied by physiological sensations similar to those experienced in the pit of the stomach while ascending in a fast elevator

11. They find, or more usually their partners or family members discover, unusual marks on their bodies such as scars, scoop marks, laser-like straight-line cuts on limbs or back, and are unable to explain how or when they had occurred

12. They sometimes wake up with bruises, some quite large and painful, or fresh burn marks, especially around wrists, upper arms or ankles—none of which they can explain. Abductees have been known to wake up from sleep with actual *fractures*, usually in their limb extremities

13. Some wake with leaves, twigs, or soil in the bed, as if they have been outside during the night and brought these back into bed with them

14. They awake with their nightclothes on backwards, or inside out, or else neatly folded on a nearby chair, or occasionally an item of jewellery or clothing will be missing and never subsequently be found. Some have woken up wearing strange clothes which do not belong to them

15. It is not uncommon for those experiencing a 'missing time' episode while awake and then going about their daily business to discover later that items of jewellery are missing or wrongly fastened. For example, one abductee discovered, upon arriving home two hours late, that her earrings were fitted with the posts and clasps reversed. Another that his shirt buttons were misaligned and his wristwatch strap was done up the wrong way round so that the watch face appeared upside down as he looked at it

16. A significant minority have found themselves on at least one occasion locked out of their homes in the middle of the night wearing their nightclothes, in some cases obliging them to resort to forced entry into their own home. One abductee I know found himself in the garden and locked out of the family home in the middle of the night on two separate occasions when he was a child

17. Many have had at least one clear UFO sighting during their lives, and sometimes several

18. They often have a compulsive fascination with UFOs from an early age, and voraciously read up on the subject. However, a minority react strongly against even the mention of the subject and will go out of their way to avoid all contact with it

19. Some report seeing flashing lights or moving balls of light in their homes, even during the daytime

20. They have an oppressive feeling of being watched and become obsessed with personal security

21. Some suffer sudden or frequently recurring illnesses, sinus problems, fatigue, fevers, migraines, or rashes which defy medical diagnosis

22. They have an aversion, sometimes overwhelming, to medical procedures or doctors

23. Many have had, on at least one occasion but often more, a powerful compulsion to walk or drive to an isolated location and wait there, without any obvious explanation or cause [8]

24. They often have aversion, amounting to a clinical phobia, of heights, snakes, spiders, large insects, certain sounds, bright lights, and fear being alone

25. Many wake up in the night, sometimes frequently, feeling a sense of panic or anxiety for no apparent reason. It is common for abductees to be unable to sleep at all in complete darkness, and they keep a low light switched on in the room

26. They experience nose bleeds during periods of their lives—particularly in childhood—and repeatedly wake up with blood on the pillow and soreness in the very back upper area of the nose, without apparent cause

27. They often suffer from sinus pain, sometimes to a serious degree and usually centred behind one eye. This can be so severe that some report total incapacity for up to thirty-six hours, prescribed painkillers offering no relief. These episodes are reported to occur rarely but at precise intervals, like once each nine or eighteen months, with no incident in between

28. Most abductees report a monotone ringing in the ears, often just in one ear, clearly and of several seconds' duration once or twice during the day

29. They often report problems forming sexual attachments or experience sexual dysfunction, despite being considered conventionally attractive

30. Some report odd feelings that they must not become involved in a relationship at all, because they feel it would "interfere with something"

31. Both men and women report that they wake up with pain in their genitals, often quite severe, which can last all day, of no obvious cause

32. Female abductees occasionally report missing pregnancies during the first trimester, with no bleeding or other evidence of the embryo being expelled

33. Polycystic ovaries and other gynaecological problems seem to be common in female abductees

34. A minority of abductees report that computers, TVs, tablets, smartphones, and other electrical or electronic apparatus seem to malfunction in their presence without explanation. Several report streetlights blinking or going out when they walk under them, or radios, television sets, and computers being affected when they walk by

35. A minority of abductees claim to sometimes know what other people are thinking and are able to 'listen' to others' thoughts. This ability seems to correlate with a recent abduction and endures for a couple of days before receding

Furthermore, it is important to understand how abductees view their situation once they become aware of it:

1. Most do not welcome abductions and want them to stop

2. Almost all (with rare exceptions) want to avoid any kind of public exposure. Fear of ridicule keeps many from seeking help or even discussing their experiences with anyone

3. High-functioning people who report abduction experiences testify against their own interests, knowing that public exposure could seriously impair or ruin their careers

It is within the framework of societal attitudes which dismiss outright this phenomenon, leading to such caution by the affected individuals, that the program continues, rarely impeded or distracted by its target population from the realisation of its objectives—whatever they may eventually prove to be.

Next, we examine what physical evidence the abductors leave behind, much of it on and inside the bodies of abductees, which lends further credence to this phenomenon being a very physical reality.

Chapter Four

Physical Evidence

What they leave behind: scars and scoops, implants and hair

That many—perhaps most—abductees find, or more usually their partners or family members discover, unusual marks on their bodies such as scars, scoop marks, laser-like straight-line cuts, and similar abnormalities on limbs, the back, or other areas of the body following abductions and are unable to explain how these occurred, is now beyond question.

Scoops and Cuts

These scars exist and have a narrow range of definite patterns. They have been filmed, photographed, and sometimes inspected by medical professionals. They almost never appear on the face or head area but are administered on parts of the body where they are least likely to be noticed by the abductee. I myself have at least two separate scoop mark scars, one on the back of the right leg just above the knee and a second, larger one, on the rear of the right shoulder. (See images in Appendix B.) The mark is deep and symmetrical but displays two small parallel horizontal ridges inside the crater, as though the two edges of a cutting instrument were coming together.

The second, larger scoop mark was first noticed by my wife following two abductions on consecutive nights, that of 23–24 August and the following night, 24–25 August 2008. Interestingly I originally thought the first might be an 'out-of-body experience' (OBE) but soon afterwards realised that it was a night-time abduction.

When visited the second night it was immediately obvious to me what was going on, as I saw very clearly the three small greys escorting me upwards through the night sky, one ahead and above, and two behind and below. One of the three escorts put the image into my mind that a young relative, my second cousin then aged six, was with us. Indeed, she was right there, less than a metre away with her long, curly blond hair, a long nightdress and big smile, in vivid technicolour. Or so I at first thought. As I stretched out my arm to her my hand went straight through the image to make contact with a cold, skinny torso, as briefly explained in Chapter One. She had obviously been a projection put into my mind by my abductors as a calming palliative image to minimise panic, distress, or resistance. As I touched it, the fear and anxiety of the small creature was so strong that I actually felt some remorse.

In my experience, abductions on consecutive nights are unusual so I have no idea what might have prompted this *except* that I could feel a viral illness (probably influenza) starting up in my respiratory system a day or so earlier, and by the morning after the second incident, all trace of it was gone. It is rare in my experience that the abductors do such things but other abductees have occasionally claimed to experience illnesses being cured. I can honestly say this is the first and only time it ever happened in my case—if, indeed, that was the cause.

Regardless, I was delivered back with a fresh, raw scoop mark scar behind my right shoulder to add to the older and smaller one on the back of my calf. We photographed these scars and sent the images to a couple of abduction researchers to find out whether they matched those found on others. It turned out that they did.

In late September 2008 Budd Hopkins contacted me by telephone, calling from his home in Manhattan. He had examined the photos of my scars, which had been shown to him by Peter Robbins.[1] He asked if I would be willing to visit him in NYC to appear in a documentary film about the abduction phenomenon. (It was in truth planned as a biopic about *his* investigation of this phenomenon and subsequent work with abductees to uncover the core aspects of the subject, and the recognised expertise he had achieved in the field.) This visit would involve on-camera examination and biopsy of the scoop-mark by a qualified, practising dermatologist. He invited me to stay at his apartment in December 2008 for a week to do this, and also to work with me on the experiences. He asked whether I might consent to these sessions being filmed. As described in Chapter Six, I had attempted to contact Hopkins without success throughout 2008, so this serendipitous call was as welcome as it was unexpected.

On Monday 8th December, my wife Janis and I arrived at JFK airport on a flight from London to be greeted by Budd Hopkins himself, whom I recognised instantly. He was accompanied by a film/sound crew who began recording as soon as we were identified. The documentary was being made by Breakthru Films whose work has won numerous awards.[2]

We all piled into a large MPV yellow cab, with the film equipment and our luggage, and off we went to Manhattan through the winter sunshine and freezing wind.

Budd lived at 246 w16th Street in Lower Manhattan,[3] a short walk from Chelsea Market and the West River. His open-plan apartment occupied an entire floor with a separate large, high-ceilinged studio at the back of the building. For a week, we stayed there and were interviewed on camera by both Budd and Ricki Stern, the film director and co-owner of Breakthru. We were filmed coming and going and doing this and that. Legal release forms were signed giving Breakthru ownership rights to the films and interviews.

As it was December, it was freezing cold outside. Whenever we left to go out we had to dress up in layers of warm clothing to brave the bone-chilling winds. It is often said that New York City has only two seasons and makes do with no Spring or Fall. It's certainly true that the Spring and Fall can be short, and the seasonal temperature change quite sudden. For months, it's hot summer then one day, the temperature drops and suddenly it's winter— same for the onset of summer in April, which comes on all at once over a few days.

The heating system in Budd's building dated from the 1950s and the furnace broke down one day during our stay, necessitating an emergency call-out to have someone fix it before we all froze to death. Remarkably, the heating engineer who showed up spoke with Budd about his abduction experiences as he fixed the furnace.

On Thursday 11th December 2008 we went with the Breakthru film and sound crew to the Ackerman Academy of Dermatopathology, at 10 E88th Street, where my scar was photographed, punch-biopsied, stitched. We then spoke with Dr Joel Kassimir and his team. A second abductee—whom I met only on this one occasion—with scoop mark scars behind both knees was also present for biopsy and analysis. A written pathology report followed some weeks later. Breakthru Films arranged, and paid for, this biopsy procedure and thus own the rights to the lab report, but the letter from Dr Kassimir to Hopkins is reproduced in Appendix A.

It turns out that the technical term for this kind of injury is *dermata fibrona*. The dermatologist's opinion was that the scar was probably not an insect bite, disease scar, or likely to be any kind of accidental injury as it was too symmetrical in form. He told us he had never seen a DF so large and evenly formed, and moreover had never known of one of any size where the patient claimed no memory of how it occurred.

A minor irony is that following Dr Kassimir's procedure the symmetry of the original scoop mark is now partially obscured by the surgical stitching. But I still have the original photographic images for comparison, which you may find in Appendix B.

The Implants

In April 2009, I was privileged to meet with Dr Roger Leir, a practising Podiatric Surgeon. Dr Leir's involvement in identifying, surgically removing, and analysing alleged implants from the bodies of serial abductees around the southern California area may be the nearest thing we have to irrefutable proof that the abductors do place small nano-devices in abductees' bodies, for purposes about which we can only speculate. His book, *The Aliens and the Scalpel*,[4] is a fascinating addition to research of the abduction phenomenon.

During the nineteen years from 1995 to his death in 2014, Dr Leir surgically retrieved eighteen foreign objects—mainly from the legs and feet—in seventeen different patients. Only one of these objects turned out to have a prosaic origin: it was a piece of broken glass. The others, however, proved to be far more mysterious: some of them at least are constructed of exotic isotopes not found on Earth, and moreover proved to be highly complex nanostructures strongly suggestive of extraterrestrial origin.

Dr Leir's original offer in 1995 to surgically remove suspected alien implants from serial abductees had been taken up by Derrell Sims [5] at a MUFON meeting. Several patients were selected. The involvement of a large number of people in planning the surgeries became necessary, from nursing assistants to anaesthetists to assistant surgeons, to someone to collect abductees from the airport and drive them to the surgery. The surgeries were professionally recorded, filmed and relayed live to an adjacent room full of invited guests, such as interested medical professionals and others who worked in bio-medical disciplines.

The nature of the objects removed during the first and second groups of surgeries were most interesting. There was no evidence of an entry wound at any of the implant sites. The surrounding tissue was completely absent any signs of inflammation which would normally be visible even to the human eye

Nick Pope & Dr Roger Leir sharing a joke, 17 April 2009.

where an unknown foreign body, however small, penetrates through the skin and lodges inside bodily tissue. Which begs the question: how could they have got into the patient?

Each of the minuscule objects was found to be placed inside a tiny, woven shell of hemosiderin, protein coagulum, and keratin. Haemosiderin, found in red blood cells, is associated with 'tissue respiration', whereby CO_2 is absorbed and O_2 simultaneously released. Protein coagulum is involved in blood clotting, working with the body's thrombocytes. The presence of these two substances may be credibly explained, though how and why they might be found together in this kind of structure is another matter.

Keratin is the core substance in hair and fingernails, and the primary constituent of the external epidermis which creates a barrier to infection from environmental pathogens. It is not found internally anywhere in the human body, excepting the hair roots which extend only into the hypodermis, the innermost layer of human skin (which happens to be twice as thick in adult females as it is in adult males).

Nerve proprioceptors protruded from each tiny structure. These are nerve cells normally found in the greatest numbers in the face and lips, the fingertips and external sex organs. What were these hypersensitive nerve receptors doing embedded in a nanostructure deep inside the bodily tissue?

Their only biological function is as external sensory receptors, so you don't need them, and none are normally found, deep inside bodily tissue.

Now, we can speculate all day about what might be going on here but one thing we can say, with certainty, is that no way do these things exhibit the characteristics normally seen with foreign bodies which find their way, through accident or other means, into the human organism. In 'normal' cases, a foreign body would generate substantial and persistent inflammatory tissue response and nerve tissue would never develop around it. It seems these things were somehow designed to be bio-compatible with the human body.

Inside each of these essentially biological shells, described as like, "cantaloupe seeds radiating small tendrils," was discovered a mainly metallic nano-device. The morphology of these devices was spheroid, or triangular, or occasionally T-shaped or a rod-like micro-cylinder.

Here's how the objects were tested in the lab. All analyses were double-blind.

1. Density immersion technique in toluene
2. Mechanical properties analysis including hardness and elastic modulus
3. x-ray energy-dispersive spectroscopy
4. Scanning electron microscopy
5. x-ray diffraction pattern analysis
6. Electromagnetic properties analysis

Other interesting characteristics included:

- All objects were recovered from the left side of the body

- All of the specimens emitted radio waves with a powerful electromagnetic field, easily detectable with a gauss meter

- When removed from the human body, all specimens fluoresced under UV light

- During laboratory analysis, the objects were found to be complex, containing a diversity of elements (Al, Ba, Ca, Cu, Fe, Mg, Mn, Na, Ni, Pb, Si, Zn) not normally found together inside tiny natural objects

- Xeropthalmia (night blindness) was common to 50% of the male patients and 90% of the females. In the general population, the percentage is in single digits: of course this might be just a bizarre coincidence but the extremely high percentages are notable

One of the samples, given the designation T1-2, turned out to have a core of iron. However, the isotopic ratios differed from those in earthly iron. They were essentially similar to those found in iron-nickel meteorites, many of which have fallen to Earth over the years: in other words, extraterrestrial in origin. Isotopes are two or more forms of the same element that contain equal numbers of protons and differing numbers of neutrons in their nuclei, and as a consequence have a different relative atomic mass while retaining the same essential chemical properties. It's odd that a micro-piece of iron with non-terrestrial characteristics managed to find its own way into an abductee's body, wouldn't you agree?

As mentioned above, all of these objects were measured to be emitting radio waves. Frequencies in the range of 8HZ, 14MHZ, and 19GHZ were found. All terrestrial objects, even in the natural world, emit some electromagnetic radiation, from long frequency radio waves to short frequency microwaves, dependent mostly on temperature. In the case of the objects retrieved by Dr Leir, the radio emissions mysteriously ceased between 60–90 days of their removal from the patient, suggesting their activity was somewhat or wholly dependent on their continuing to be sited *inside* the human body. Who would have implanted them, and how? More pertinently, to what purpose?

Robert Bigelow, an entrepreneur who has funded the National Institute of Discovery Science (NIDS), and Dr John Alexander, a retired US Army Colonel who has taken a great interest in UFOs and related controversies, invited Dr Leir and Mr. Sims to visit NIDS in Nevada. The NIDS board eventually offered to finance their research, including further surgeries. They would also produce Dr Leir's book, *The Aliens and the Scalpel*. This work would lead to Leir being invited to conferences all over the world to present his findings, though he eventually came to tire of the exposure. Roger Leir, with whom I corresponded sporadically until 2012, died on 14 March 2014, four days before his 80th birthday. RIP

It must be stated that, because Roger was a specialist podiatric surgeon, the implants that he removed had all been extracted from the extremities, especially the feet. Many abductees around the world do claim to have implants in these areas, but the more interesting sites are in the head, especially deep in the ear cavities from where most abductees experience a distinctive ringing monotone of several seconds' duration often one or more times each day. An even more common site is at the top of the nasal passages above or next to the sinus cavities, placed near the olfactory nerve as it passes through the cribiform plate, where it enters the brain. These often cause serious problems for abductees including prolific nose bleeds and intolerably painful sinus headaches, normally centred behind one eye, which can be of thirty-six hours' continuous duration, and do not generally respond to pharmaceutical painkillers, making it impossible for the abductee to engage in any normal activities for the duration.

I have experienced all of these. The monotone ringing is in both ears, in my case a higher tone in the left eardrum and a lower tone in the right. The right one rings almost every day and sometimes twice in a day; the left one rings less often, maybe once in a month. I have no idea what this signifies, but the hearing in the right ear has become impaired—though this might in my case be partly genetic in origin: my late father suffered from poor hearing and late in life became increasingly deaf.

The nose bleeds I suffered regularly as a child, especially between the ages of seven and nine years old, and medical attention was sought at the request of my school, as the blood often started to flow during class.

The sinus headaches I suffered regularly, exactly every eighteen months almost to the day, in May and November, for twenty years. The last one was in November 1992, when I was 36 years old, from which date they stopped and never resumed. I have suspected for years, but of course could not prove, that either the implant had been removed and not replaced, or else may have been replaced with a less intrusive and troublesome device. Or, perhaps the taller grey alien 'doctor' who had me as part of his assigned workload for twenty years was replaced by another whose surgical technique in placing these devices differed sufficiently to cause no pain.

I was given confirmatory evidence in late February of 2022 that the device was in fact replaced, during an incident related below, though it's possible the more recent implant/s are in a different site: the top of the nasal pharynx rather than the sinus cavity.

Steve Colbern—not to be confused with the satirical chat show host Stephen Colbert, but a chemist/materials scientist with more than twenty years' industry experience—wrote a technical paper in 2009 on one of the samples removed through Dr Leir's surgical intervention. The sample was designated Number 15 and titled, "Analysis of Object Taken from Patient John Smith" (a pseudonym).

Robert Hastings, the author of UFOs and Nukes, has a professional background with scanning electron microscopy and elemental analysis, acquired during his 14-year employment as a laboratory analyst with Philips Semiconductors. He spoke with Colbern in July 2019 to discuss his findings, recording these exchanges in detail in his 2019 book, *Confession*. Below is a brief summary of the main takeaways of the exchange:

1. While analysing the composition of the apparently artificial object via Inductively-Coupled Plasma Mass Spectrometry, Colbern found isotope ratios in four elements—boron, magnesium, nickel and copper—that are not found on Earth. Whomever created it had to have used one or more extraterrestrial sources for those elements in its manufacture. Also detected were trace elements such as iridium, gallium and germanium, which are rare on Earth but abundant in iron-nickel meteorites.

2. Using a scanning electron microscope, Colbern had found what appeared to be nano-scale artificial structures, termed 'carbon nanotubes', suggestive of electrical current-carrying pathways. (A nanometre is one billionth of a metre.)

3. One independent source says: "A carbon nanotube is a honeycomb lattice rolled into a cylinder ... one of the most significant properties of carbon nanotubes is their electronic structure which depends only on their geometry, and is unique to solid state physics. Specifically, it is either metallic or semiconducting ... thus we can imagine that the smallest possible semiconductor devices are likely to be based on carbon nanotubes."[6]

4. Other, horn-shaped, nano-scale structures present in the sample, Colborn speculated, might be antennae. Radio waves—in the 1.2GHz, 110 and 17MHz, and 8Hz bands—were in fact detected in the immediate region of the object from the patient's body, indicating that it had been transmitting a signal.

5. When I asked Colbern if any of the radio frequencies were associated with human applications, he said that the 1.2GHz wavelength band is used for communicating with satellites, because it is not easily absorbed by the atmosphere. My response was, "So we use that band to communicate with satellites. I wonder if someone else uses it to communicate with other types of craft orbiting Earth." This of course was speculative on my part.

6. Summing up his data, Colbern says: "The extreme difference in the isotopic ratios of the sample elements from the isotopic ratios of elements found on Earth provide strong confirmation that the material in the sample is of extraterrestrial origin."

7. The structures of the non-metallic portion have unusual shapes which suggest artificiality and functionality. This, along with the fact that the object was giving off radio signals before removal, strongly indicates that this is a manufactured, nanotechnological device ... inserted in patient Smith for a definite purpose ... the function of the device cannot be determined with certainty from the available data and the device may have had multiple functions and missions.

8. The manufacture of a device comparable to this one is probably beyond the technology of known, Earthly, commercial processes, at the present time. It is most likely, therefore, that the device was manufactured by an alien civilization. It is still a possibility, however, that the device was manufactured by some process known to the Earthly military/Black Project community.

Colbern's full report is available online [7] and is recommended reading for those who understand the technical complexity of semiconductors and related technologies. A significant abstract of the report is reproduced in full in Appendix D, together with the web address if you want to read the original online.

Hastings' book, *Confession*, which was co-written with former USAF officer Dr Bob Jacobs—who was central to the Big Sur Incident of 1964[8]—is also highly recommended reading. It is referenced in the bibliography and is available in print or as an e-book.

Scientific Conference Presentation at MIT

On 13–17 June 1992, the Abduction Study Conference was held at MIT, an event attended by some 80 specialists studying the phenomenon. Among them were medical practitioners, psychologists, qualified experts in the various STEM disciplines, abduction researchers, hypnotherapists, and abductees.

The conference was co-chaired by Dr John E. Mack, of Harvard Medical School and Dr David Pritchard, Professor of Physics at MIT. Dr Mack founded the psychiatry unit at Cambridge Hospital, in addition to being the founding director of the Center for Psychology and Social Change—renamed after his passing as the John E. Mack Institute. Dr Pritchard is a National Science Foundation Pre-Doctoral Fellow, and a Fellow of both the American Association for the Advancement of Science and the American Academy of Arts and Sciences. He was the recipient of the 1991 Herbert P. Broida Prize for chemical physics.

The Olfactory Nerve: Biopsies and Implants

Dr John D. Miller, MD, FACEP, (Fellow of the American College of Emergency Physicians) is a partner in the Southern California Permanent Medical Group, and practises emergency medicine specialising in neurosurgery cases.

During the 1992 MIT conference, Dr Miller offered some interesting insights into why the olfactory nerve passing through the olfactory mucous membrane might be chosen by the abductors as a useful location to lodge one of the small implants, as it could possibly monitor certain brain activity while being very difficult to access, remove, or even detect by human surgical intervention.

> Alien doctors are often reported to probe abductees' nasal passages with sharp objects. Interestingly they are not reported to use a speculum to examine the nose before or during this blind probing … human doctors do not blindly insert sharp objects up our patients' noses.
>
> The abductee often reports a painful crunching sensation and a feeling that the brain was penetrated during this alleged alien nasal probing. Later a nosebleed may be experienced.

I offer a bit of anatomically based speculation on this: The nasal cavity, aside from being a possible "parking place" for the spherical implants that are sometimes reported, has one unique structure that could be of special importance far up in the deep recesses of the nasal passages and well behind the external nose. This is the OLFACTORY NERVE (the first cranial nerve—the nerve of smell). This alleged alien blind nasal probing could represent a biopsy of the olfactory mucous membrane, which has special features that should be noted.

Anatomists classify the olfactory nerve as a fiber tract of the brain (Chusid 1985:110–112). The cribriform plate of the ethmoid bone is a thin sieve-like bone with many tiny holes in it. It separates the upper nasal passages from the bottom of the frontal area of the brain. Through the tiny holes in the cribriform plate pass the thin fibers of the primary neurons of the olfactory nerve. **The cell bodies and nerve fibers of the primary neurons are found in rich abundance in the olfactory mucous membrane in the upper nasal cavity but outside the cranial vault.**

Apparently these are unique cells in that they are the only example in the human of peripherally placed sensory ganglion cells similar to those found in certain lower animals (Copenhaver et al 1978:784–785). The olfactory mucous membrane itself is restricted in location to a small area just below the cribriform plate. This area cannot be visually inspected without special instruments and experience. The olfactory mucous membrane is not widely distributed in the nose as one might suspect (Netter and Colacino consulting ed. 1989 plate 38).

The removal of some of these neurons will not cause the person any detectable neurologic deficit. Function of the olfactory nerve can only be very grossly assessed by a human examiner.

Thus the primary olfactory neurons in the olfactory mucous membrane represent a site where small amounts of central-type nervous tissue could be surreptitiously harvested. Could this be the purpose of these strange nasal probing procedures? Remember that some abductees offer subjective complaints of problems with their sense of smell.

Given the overall apparent focus of the aliens on the cranium and neurologic function I see this as a possibility but note that this is mere speculation.

Olfactory nerve biopsies by human doctors are very rare but our neurosurgeons do sometimes operate in the area of the cribiform plate, traumatic perforation of which is a very serious and often ultimately fatal injury.[9]

My Own Personal Nasal Implant "Misfortune"

I have my own personal story about an upper nasal implant. As mentioned earlier, I had long been suffering from debilitating headaches which struck regularly for twenty years, then disappeared.

On Saturday 26th February 2022, I sent the following email to Robert Hastings and Bob Jacobs:

> I am fairly certain I coughed/sneezed out an implant, possibly sited in the olfactory mucous membrane (just outside the cribiform plate of the ethmoid bone which separates the upper nasal passages from the bottom of the frontal area of the brain, through which passes the olfactory nerve). I have a history of 'sinus headaches' of 36 hours' duration exactly every 18 months and have long suspected that these coincide with the re-siting of an implant in precisely this location.
>
> The past week I had a head cold, and a couple of really hard sneezes yesterday may have dislodged it. Yesterday evening I felt something loose, right at the top of my nasal cavity and inadvertently snorted it down into the trachea. A minute later I coughed up a dark, clotted bloody mass into the wash basin.
>
> Now here's the weird part: I was very anxious to flush it away and get rid of it, and only felt the anxiety subside when it had gone, flushed by a gallon or so of water from the tap.
>
> This was followed by a steady nosebleed from both nostrils for a couple of hours. Sorry to be so anatomical, but to find out how far up the blood was coming from I rolled a long piece of tissue up into a narrow tube and inserted it right up into the very top of the nasal cavity, same on the other nostril. That site is where the injury is, and where the clotted mass obviously came from.
>
> Result: we do not have an implant to examine, as it's flushed into the septic tank. But I am confident that one was sited there and is now expelled. They'll probably replace it in due course.
>
> No need to worry as the bleeding stopped within a couple of hours. Abductees often complain they gradually lose their sense of smell, and the siting of such implants is probably why.

In fact, the nosebleed following the expulsion of this suspected nasal implant lasted more than a couple of hours before completely subsiding.

Though the prolific bleed lasted only two hours and then reduced in volume, it was a full 24 hours before the blood flow stopped altogether, so the injury caused by the expulsion of the implant was obviously quite deep.

My energy levels were unexpectedly and exponentially increased once it was out, and my wife tells me the resultant boost in vigour is so notable that I seem to have suddenly become ten years younger. I certainly feel that. Goodness knows what that implant thing was doing.

I wish I'd been smart enough to realise what it was lying there in the wash basin but was seized with a great urgency to wash it down the drain and it's now in our septic tank. Where, presumably, our little friends will continue to monitor its output. We have unfortunately lost the opportunity to analyse a valuable implant. My panic reaction is commonly reported on the rare occasions that bodily implants are unexpectedly expelled by abductees, so perhaps some strong suggestion is planted by the abductors to insure against the possibility of discovery, leading to analysis and consequent exposure of their nano-biotechnology.

Postscript to the Incident

We have a three-year-old English Rough Collie, a household pet who acts as a guard dog for us in the remote country location where we live. He's good at night, content with his life with us and normally sleeps soundly for ten hours each night on the floor of my studio, or very occasionally on the bedroom floor.

On the night of 3–4 April 2022, our Collie was unusually agitated. My wife eventually got him to settle down, as I had retired early, and she followed shortly afterwards when the dog was finally calmed.

In the early hours, I awoke feeling unusually agitated. I suffer from chronic hypertension (thankfully well-managed these days) and, on checking my blood pressure discovered it to be unusually high. A small quantity of blood was coming from my right nostril. It continued to seep for the following twenty-four hours.

It seems 'our little friends' may have re-sited the implant, having noted after six weeks that it had stopped transmitting. As mentioned above, Roger Leir, Steve Colbern, and others have discovered that once removed from the body, the devices cease transmitting after several weeks.

Due to the parlous state of public discourse on the subject, especially among the professional medical community, it's obviously a non-starter to visit the family doctor and request an x-ray for a possible alien nasal implant. Its function is unknown and likely to remain so, because they don't tell you. Abductees are basically on their own in being left to cope with this thing. It's no fun, I can tell you.

Some Other Bodily Systems of Interest, and Some Obviously Not

Before we leave Dr Miller, he has some thoughts on the procedures reported to be carried out by the alien 'doctors'—the taller greys. This part of his lecture does not concern implants, but is interesting background nevertheless.

> The physical exam seems to consistently focus on certain systems and omit or skimp on other systems. The systems that are carefully examined are the cranium and nervous system, skin, reproductive system and at times the joints. The exam seems to omit or shortchange the cardiovascular system, the respiratory system below the pharynx, and the lymphatic system.
> ... The thorax and abdomen above the umbilicus are given little attention in most of these reported alien exams. The cardiac and pulmonary exam, a mainstay of human physical examinations and of great concern to the human patient and physician, seems absent or at least not clearly identifiable in most reports. Sometimes a witness does report a device being placed on his chest, but I can't identify this as an EKG, chest x-ray or echocardiogram in progress.
> A human doctor when examining the abdomen will give equal attention to all four abdominal quadrants. To me, the aliens seem to most often shortchange the upper abdomen in favour of the lower. They seem for the most part to be unconcerned with the upper abdominal contents (liver, spleen, stomach, pancreas) which are of great concern for the human physician.
> If any sort of lymphatic exam is done it is hard to hear it in abduction accounts. I have not heard of aliens palpating the axilla, the site where we look for abnormal lymph nodes.
> The cranium is a great focal point of the aliens' exam but their techniques are strange. We human doctors don't generally stand at the periphery of our

patients' visual field and stare at them. We have no mindscan procedures, we have to ask questions.

When a witness told me the beings put certain strangely shaped devices in her ears and told her telepathically that she would hear "the voice of God" I knew I wasn't hearing about my kind of medicine. When she related that the beings became agitated when she refused to believe this, I was totally lost.

Human doctors don't completely remove a patient's eye and then put it back again (with the eye functioning normally!) as this has been reported in some of these stories.[10]

Although this horrific procedure is outside my personal experience or memory, I can confirm that this has been reported by several abductees, including Betty Andreasson and Ted Rice. Ted was the sole subject of Dr Karla Turner's third book on the abduction phenomenon, *Masquerade of Angels*. I corresponded with Ted for two years: his experiences have been lifelong and very complicated. According to the testimonies of Betty, Ted, and others, what they allegedly are doing is siting an implant in the brain of the abductee where it will be inaccessible to human surgical probing and so cannot be removed or interfered with.

As some systems are seemingly shortchanged by the aliens, the skin (dermatologic) exam seems exaggerated. Aliens are often reported to inspect the entire skin surface minutely. Additionally they are reported to become startled or agitated when they find scars or new marks. The human doctor often gives the skin only a superficial glance unless the patient is seriously ill or has a skin-related complaint. If abnormalities are found they don't cause agitation on the part of the doctor.

Although female witnesses often report "gynecologic" type exams by the aliens, I don't recall ever hearing a witness report a bimanual pelvic exam, the absolute mainstay of the human gynecologic exam. This is done by the human doctor by placing one or two gloved fingers inside the vagina and palpating above the pubic bone with the other hand in order to feel the pelvic contents. The absence of the bimanual exam in the course of otherwise apparently extensive gynecologic manipulation is puzzling.

Another striking absence is the digital rectal exam. The beings are reported to insert various, sometimes quite large, devices into the rectum but do not precede this mechanical insertion with an exam by a gloved digit. Human doctors are taught to perform a digital exam before inserting any large object (such as a scope) into the rectum.

It is notable that the alien doctors are never reported to wear surgical gloves [nor any other kind of PPE].[11]

Hair of the Alien

I have another item of hard evidence to confirm the reality of the abduction phenomenon, recounted below. Readers are cautioned that this one is super-strange, and it happened in 2015 during an abduction in my Hertfordshire house.

First of all to set the scene, some facts about a 1992 event in Australia involving a man called Peter Khoury.

In 1992, 28-year-old Peter Khoury, who had a history of abduction encounters, reported an extraordinary experience in Sydney, Australia. This specific incident happened on Thursday 23rd July at around 7:30am, according to Khoury's diary and the testimony of his wife Vivian to whom he related the story of his encounter three weeks after the event.

Khoury was a first-generation immigrant to Australia from Lebanon. At the time of the incident he was recovering from injuries sustained after being violently assaulted with a shovel on a Sydney building site at which he had been working. That morning in July 1992 he drove his wife to the railway station for her journey to work, returned to the house around 7:05am, then and went straight back to sleep to continue his convalescence.

What followed was a bizarre encounter with two humanoid females whom Khoury discovered when he awakened from a short sleep to both be kneeling on the double bed in which he had been up to that point sleeping. He describes one of the uninvited visitors as blond-haired and the other as "oriental-looking." Both were described as having unnaturally long heads, pointy chins "like a witch's long chin in the movies," and "eyes two to three times bigger than our eyes." The oriental-looking one had unnaturally prominent cheekbones, "like she had been punched in the cheeks by Mike Tyson or something." Both beings appeared to be naked.[12]

Excepting the abnormally long heads, chiselled faces and large eyes, Khoury reports they otherwise looked like normal human females—except that the blond one seemed to be very tall. Peter Khoury is 182cm tall, and reports that "when she sat up ... I would say she had a head and a half higher than me." Of course, it's quite possible that Khoury overestimated the height of his visitor, as he was lying down and at no time did he recall either of the visitors ever actually standing on the floor where it might have been easier for him to assess their height more accurately.[13]

There was no sound or any type of communication with Khoury throughout the encounter but he was certain the two were "silently communicating with each other ... they knew what they were doing ... they were there for a reason ... the first thing that came into my mind was 'babies.'"[14]

You may read Peter Khoury's full account of this incident in Chapter Two of Bill Chalker's book, *Hair of the Alien*, which not only describes this encounter in detail but reveals that Khoury was almost certainly an abductee in the program, with a life-long history of strange events. One extraordinary incident is described by his mother as having occurred in Lebanon when Peter was a baby.

During the July 1992 encounter, Khoury claimed the blond alien forced his head onto her breast, to which he reacted by biting her nipple: "I have no idea why I did that," Khoury later admitted, "but I think that it was the only way I could say that I didn't want to do this ... she was pretty strong; when I'd resist, she would pull me back with ease." This bizarre reaction left him with a small piece of the bitten-off nipple lodged in his throat, causing incessant coughing. He reported that it was three days before his throat cleared and he either finally dislodged it or else it was swallowed or dissolved. "There was no blood, there was nothing, no trace whatsoever. It was as if I took a bit out of a plastic dummy or mannequin that was made of rubber or something. Maybe I was a fool to bite the breast ... you handle situations on the spur of the moment. Maybe I should have done it differently. But that was my way of dealing with it, to put a stop to it there and then ... it was so clinical ... they had no emotion ..." While all this was happening the Oriental-looking one just watched passively while crouched on the mattress with her feet tucked-in under her hips, as if she was "there to learn how to do something."[15]

As Peter was overcome by a coughing fit, the visitors "performed a vanishing act and suddenly they weren't there anymore." He quickly got up out of the bed with an urge to visit the bathroom, where he discovered two blond hairs entwined around his penis. According to Khoury, the entwined hairs caused extreme pain and it took some time to disentangle them. One of the hairs was 10–12cm in length, the other 6–8cm and the two hairs looked otherwise identical apart from the differing lengths. He deposited them in a sealed poly bag.

Khoury was understandably reluctant to mention his encounter to his wife. Embarrassed by the incident, he did nothing with the hair samples until he found them in 1999, following his meeting UFO and abduction investigator Bill Chalker. Khoury had sought out Chalker because he had investigated the famous August 1993 Kelly Cahill abduction near Melbourne which, exceptionally, "was observed by three separate groups of independent witnesses and supported by an intriguing array of physical evidence."[16]

Forensic DNA analysis is not a single technique, but a range of laboratory tests which have been coaxed out of the vast expanse of the human genome. Most of the forensic techniques are generally powered by polymerase chain reaction. PCR is a process which allows the reproduction of very small amounts of DNA, making possible simultaneous detailed analyses of the genome by multiple laboratories in different locations. PCR was developed by Dr Kary Mullis, the only Nobel Prize laureate to openly admit a possible personal alien abduction experience, which is related in some detail in his autobiographical work, *Dancing Naked in the Mind Field*.

The DNA analysis of one of the two "exceptionally thin hairs," as organised by Bill Chalker revealed some intriguing results.[17] The blond hair has Basque-Gaelic DNA in its root, but Chinese DNA in two other parts of its shaft which in 'normal' circumstances is impossible and indicative of highly advanced genetic engineering. Furthermore, the PCR bands imply that the blond alien might be homozygous for the delta-32 mutation of the CCR5 gene,[18] so bred to be genetically resistant to viruses as diverse as HIV and smallpox.[19]

More Hairs of the Alien(s)?

As mentioned in an earlier chapter, after an abduction in July 2015, we had discovered some hairs on the living room rug in my Hertfordshire house. A brief background might be helpful to set the scene.

The ground floor of the house was of open-plan design: a lounge area with easy chairs and coffee table, dining area with oak table and six dining chairs, and kitchen with fitted oak units were all in one big L-shaped space. The floors were of oak with a large, patterned rug placed under the oak dining table on the polished oak floor. The kitchen area floor was tiled. The smooth tiled or polished oak floors were easy to clean and were vacuumed every couple of days, and at this time we had just carried out a thorough house clean. At this time we kept no house pets and had entertained no visitors—none by invitation anyway—during the preceding few weeks.

Late one evening, between 11:30pm and midnight, I was sitting alone—my wife had gone to bed—in one of the easy chairs reading when suddenly I was overcome with a euphoric drowsiness. I should have recognised the signs, but the drowsiness was so overpoweringly pleasant it was almost irresistible. Before consciousness left me, I became dimly aware of two lean figures entering the room through the full-length glass doors leading to the enclosed back garden. That is, they entered *through the glass* without opening the doors.

While fighting to stay awake, the visual impression received through my increasingly immobilised state was that the leading and slightly taller figure was female and the other slightly shorter one, behind and to her right, was male. The male one looked like a normal young human of around 20 and was dressed in contemporary casual clothes and casual shoes. The female one looked mostly human but had dark, almond-shaped eyes with no visible whites, whose size was almost in proportion to those in a normal human face, maybe slightly larger, but nowhere near as large or dominant as those of the grey aliens. She would certainly not pass as a human at the local bus stop or on the high street, though she might just if she were wearing dark glasses and long sleeves, seen briefly from a distance. She appeared to be wearing a long, loose-fitting, light-coloured gown and had rather unkempt white-ish hair cut above the shoulders. Both figures had lean physiques, verging on skinny. There may have been more than two involved, but if so I have no memory of them.

When I awoke it was around 1:30. I went straight upstairs to bed and forgot all about the visitors.

The following day, my wife found a 'blond' hair—or so we had thought—25cm or so in length, under the dining room table on the otherwise pristine patterned rug. This hair clearly did not belong to either of us. My wife is a red-haired Irish woman and my hair is very short. We had had no visitors since cleaning the house the previous day, and in any case had no visitors in recent memory who might have had hair of such a colour and length.

Later the same day a second hair, seemingly identical to the first, was discovered on the garden patio next to the flower beds, right outside the glass doors to the garden. On close examination it looked as though the two identical hairs were bleached and colourless, rather than blond, although it is difficult to identify the precise colouration of single light-coloured hairs in isolation unless using a microscope, which we did not own.

Hybrid visitor wearing wig who left the hairs behind in 2015.

At this point I dimly remembered the two figures who'd entered the room before I became unconscious the previous evening. It might seem odd that this encounter could be so casually forgotten, but it's just this kind of amnesia which is common to abductions.

We put both hair samples in a sealed polythene zip-pouch and stored them in a secure place. My wife insisted that I should not know where they were stored. We then spent several months trying to find someone with the skillset, facilities, and equipment to analyse them. We drew a blank until Dave Jacobs told me on the phone that he had recently been given a hair by another abductee who remembered exactly how, where, and when she had acquired it. Dave offered to have both samples analysed by some academic colleagues in Illinois who were qualified research microscopists working in forensic analysis. Assuming the hairs were probably from a head of growing hair, there was some disappointment that neither example ended in a root-bulb, where any genetic material might have been more easily available for analysis. The laboratory analysis report may be found in Appendix C.

Both hairs—that from the other abductee 'Betsy' and that from our carpet—were found to be almost certainly *from wigs!* The dark hair discovered by the American abductee 'Betsy' (a pseudonym) was of polypropylene, commonly used in cheap wigs. The hair found on our rug turned out to be a real human hair but bleached, almost certainly from a wig because wig hairs are the only ones so treated with such chemicals, prior to the dyeing process. The wig origin also explains why neither of the hairs—plus the identical hair found on the garden patio which we retained—had any evidence of a root-bulb present.

Neither Dave Jacobs, nor I, nor anyone else we have consulted can make any sense of this, but there it is: you examine the evidence and retain an open mind. In both these encounters from 2015–16, separated by thousands of miles of ocean and reported by two different people who have never met each other, the abductors were in each case apparently wearing wigs—though wigs apparently acquired from different sources.

There was unfortunately no complex and previously unknown organic structure as revealed by the PCR–DNA testing of the two hairs Peter Khoury had found braided around a part of his male anatomy in Sydney in 1992. But the separate wig hairs analysed in 2016 revealed something almost as bizarre about the abduction program, although hardly as exotic.

This is a new detail about the program, to my certain knowledge never previously discovered. To admit this discovery was unexpected is to understate the matter, but where this knowledge might lead remains at present in the vast territory of speculation, of which this field of study has no shortage. What may be deduced with some confidence is that the wigs were acquired from human sources (a wig factory or more likely a theatrical supplies shop) and delivered elsewhere for their intended use, possibly—and here we must speculate—to assist alien-human hybrids with sparse or uneven hair growth in more effectively disguising themselves if deemed necessary when they move around in human society. Pure speculation, of course, but plausible considering the evidence.

It would admittedly be easy for a debunker to claim that this is some kind of hoax. No exotic, complex genetic material is revealed by the forensic analysis as in the Khoury case brought to public attention by Bill Chalker. You can buy a wig anywhere, a debunker would say. Of course, we all know that's true.

But that's not the point. The important thing is exactly how these hairs appeared in my home, following an abduction event. And then a completely different wig-hair was discovered by a completely different abductee somewhere in the USA, again after an abduction. I do not know this person, and she does not know me. If you don't think we might learn something from this about the program and how it's unfolding, then you're not paying attention.

"Walking Among Us"? Looks like it.

Now it's time to look at what the evidence seems to indicate might be behind the century-long abduction program: the genetic engineering of a hybrid race, physiologically identical to their terrestrial human progenitors but equipped with deeply buried neurological abilities to manipulate and control us. Scary stuff: prepare yourself for one of those 'Felix Frankfurter Moments'.

We'll also look briefly into what 'The Government' may know about what is going on, from personal experience of actions which seem to reveal an 'interest in my interest'.

Chapter Five

They're Doing *What?*

Puzzlin' evidence: get used to it ...

In 1947 President Harry Truman was reportedly briefed by senior military officers about the UFO issue. The astonishing performance of these vehicles included stationary hovering, then immediately taking off at several thousand miles per hour with no apparent period of acceleration; then changing direction so suddenly that no human occupant could possibly tolerate the resulting g-forces. When engaged by the US Air Force's first generation of jet fighters, the performance of these craft led to the conclusion that they were probably of extraterrestrial origin. The POTUS was alleged to have responded to this assessment by asking: "So, what do the sons of bitches want?" [1] Those who knew Harry Truman and were familiar with his idiomatic way of speaking found this reported response both typical in phrasing and credible in content.

A New Phenomenon: Visitors From the Stars?

First and foremost, when addressing the question about "what the sons of bitches (may) want," we have to confront and accept an obvious truth: They obviously *don't want us to know their intentions*, whatever they may be, or they would by now have found a way of communicating them to us. The only reasonable inference which may be drawn from the evidence at hand is that their intentions do not appear to be to our benefit.

These visitors may all originate from a single habitable exoplanet, though may have spread out as they developed and be now largely independent of any single planetary ecology. Perhaps those piloting these craft hail from many different places and have differing agendas. It may be that we could little comprehend what it is that some of them are here for. But, if any of them considered it important to remove doubt and uncertainty as to their intentions in the human population of Earth, then they would surely not behave as they do. Their behaviour tells us they will only allow interaction on *their* terms, at a time and circumstance of *their* choosing, when *they* can maintain total control over *all* aspects of the interaction. Questions from the natives? Not welcome.

Many people have reported interactions with the UFO occupants. The majority are abductees, who from the earliest reports in the 1950s and 1960s (some prominent early examples are described in Chapter Two) describe the encounters as being definitely on the terms of the visitors, and at a time and circumstance of their choosing. There are always outliers, such as New Agers who claim to be in possession of special knowledge or 'messages for humanity' from the aliens,[2] but as lone voices—invariably deluded narcissists of one stripe or another, however well-intentioned they might appear to be—they may be largely disregarded against the tens of thousands of abductees who admit to only puzzlement, and experience trauma, due to what they are going through as 'the visitors' choose to interact with them.

Most abductees do not seek public exposure and, due to the social stigma surrounding this subject, risk career and reputation should they go public, so the vast majority remain silent or else confide only in an investigator whose confidentiality they feel they can trust absolutely. Many refuse even to confide in their husband/wife/partner for fear they will be judged and their admission will jeopardise the relationship.

The Early Years of Public Awareness

When the abduction phenomenon began to be reported in the 1960s and 1970s it was at first assumed that the presumed extraterrestrials were simply studying the human race, capturing individuals more or less at random. Catch and release, in other words, just as human biologists have conducted their own research on animals across our planet. This paradigm held for many years.

The UFO phenomenon has been confusing enough, with sightings now in the tens of thousands. Despite a great number of excellent reports—many buttressed by photographs, film, and radar—their maddening variety and inscrutable behaviour does little to aid in assessing their intent. They didn't, and don't, 'make contact'. They still don't 'land on the White House lawn'. They have often seemed indifferent to their obviously technologically superior vessels being observed and tracked, but evidently were completely uninterested in having any dialogue with us, even on unequal terms as portrayed by the all-powerful Klaatu in the 1951 sci-fi movie, *The Day the Earth Stood Still*.[3]

From 1947, rumours of 'crash retrievals' have multiplied. Stories abound of exotic craft spirited away by the military under cover of darkness to remote, ultra-secret locations, like the legendary Area 51 in Nevada. Scores of books have been written and film documentaries made about 'retrieved UFOs', chief among them the Roswell NM incident of July 1947. But the list has grown considerably and shows no signs of abating, despite the complete lack of indisputable proof in the way of hard artefacts: just a growing list of self-declared 'insiders' who can never offer proof of anything more substantial than a desert rumour.

During the last century, retired US Marine Corps aviator Major Donald Keyhoe maintained a decades-long effort to engage the public about the UFO phenomenon.[4] Over the course of five best-selling books, and through his tireless advocacy, he asserted that the 'flying saucers' were not only extraterrestrial, but that they were conducting surveillance of our planet. Keyhoe achieved nationwide fame after publishing a landmark article in the January 1950 edition of *True Magazine* titled, "Flying Saucers are Real," which was read by millions of Americans and generated huge public interest. He spent years demanding congressional investigations into the phenomenon, charging that the government was covering up the truth by marginalising and ridiculing those who reported sightings. During a notorious January 1958 appearance on a live CBS TV broadcast, his microphone was quickly silenced when he veered suddenly from the agreed upon script. The network subsequently pleaded that they'd had no choice, due to "national security reasons." Keyhoe always maintained that the US Air Force was the principal culprit in the "Saucer Cover-up."

Many hundreds of people have investigated alleged UFO sightings, and loose networks of amateur investigators have set up all over the globe. Prominent in the USA was Keyhoe's National Investigations Committee on Aerial Phenomena (NICAP). Jim and Coral Lorenzen's Aerial Phenomena Research Organization (APRO) published one of the earliest works on the abduction phenomenon.[5] The Mutual UFO Network (MUFON) was organised on a state-by-state basis and has lasted for decades. Similar networks grew up around the world, like the British BUFORA. The UK-produced magazine, *Flying Saucer Review* became an internationally respected publication popular throughout the English-speaking world and beyond.

The French computer scientist and writer Jacques Vallée,[6] who relocated from Paris to San Francisco with his wife and young children in the 1960s, wrote that the most striking aspect of the UFO phenomenon was that it seemed to exhibit both physically real and 'psychic' aspects. In this sense, according to Vallée, it has displayed similarities with traditional folklore. He extended this idea in his 1969 work, *Passport to Magonia*. In this book, Vallée drew parallels between the UFO/alien contact phenomenon and European folklore, mainly from the British Isles, with tales of missing periods of time, interactions (often romantic or sexual in nature) between humans and faery folk from parallel 'otherworlds', and changeling babies delivered to unsuspecting mothers by the same faery folk.

Vallée was represented in Steven Spielberg's 1977 film *Close Encounters of the Third Kind* as a character named Claude Lacombe, a French scientist, played by film director Francois Truffaut.

Having read all twelve of Vallée's works on the UFO subject, many of them more than once, it eventually became obvious to this reader that although Vallée is a fine and literate writer—and he is definitely those things—he was clearly barking up the wrong tree. He repeatedly dismisses the abduction phenomenon, despite investigating and documenting several classic cases like the 1965 Valensole incident. This alien abduction included a daytime UFO landing, small three-foot-tall grey aliens with large heads and big almond-shaped eyes who paralysed the witness, and a period of missing time. Vallée's 1990 book, *Confrontations* is full of such cases, including many from Brazil, a country he visited with his wife Janine in the late 1980s. But still he refused to recognise the patterns in abduction reports, as they failed to conform to his *Magonia* ideas, though later his view moderated somewhat when faced with a ubiquity of reports of essentially similar interactions.

In his field investigations, Vallée generally accepted the memories of sole witnesses at face value, a notoriously unreliable methodology when such reports are often replete with confabulation and implanted screen memories. Vallée chose to adopt one of the classic debunkers' tired old tropes that abduction accounts result in the main from ideas cooked up by hypnotists, who place suggestions in abductees' minds that confirm their own preconceived notions.

It dawned on people only very slowly that the abduction reports offer the key to unlock the door behind which lie the purposes of the ubiquitous UFO phenomenon. Looking at and filming the airborne craft in flight tells us nothing about the purposes and intentions of these visitors, except their worldwide ubiquity: these reports of interactions with the occupants, when carefully examined, enable us to get inside the UFOs.

Common Themes Emerge

From the earliest reports in the 1960s of what we would now term *UFO abductions*, certain common themes have been discernible:

1. The descriptions of the abductors were all recognisably similar, with an occasional outlier like the Pascagoula MS case where the witnesses' descriptions of the entities differed from the norm, but the specific details of the medical procedures the two abductees reported were carried out on them did not substantially vary from those reported in other cases

2. The abducted were evidently specifically targeted in a carefully planned way and did not generally seem to be just 'taken' at random. (A possible exception might be Travis Walton, whose case is an outlier in some respects in that his 'abduction' may simply have been an act of mercy due to accident, but whose multiple witnesses makes it, paradoxically, one of the strongest.)

3. The abductors appeared to share a common methodology, in that they subjected the human abductee to seemingly specific procedures. This was a recognizable medical examination which differed markedly from human medicine in its emphases. The abducting entities seemed to know exactly what they were doing and had the tools and equipment designed specifically for use on human bodies. The way these procedures were described they looked planned, well-rehearsed and purposeful

4. A period of missing time was noticed by the abducted, of which they could recall little to nothing. To some this added a magical or supernatural aspect to the phenomenon, but to others it looked like intentional, neurological manipulation by beings who thoroughly understood how to control us. The Hill case demonstrated that hypnosis, when competently managed by a skilled practitioner, greatly assisted the recovery of memories of what had occurred during these incidents. Deliberately engineered memory blocking appears to be the norm

5. From the earliest cases, tissue sampling has been reported, almost always including sperm and ova harvesting. What the heck was that about? What would "aliens" from another star system, presumably of a totally different biology, want with all these samples? If you sample one, or two, or three human subjects in this way, you have everything you need to know about their biology and the way they reproduce. Then why was it such a persistent and repeated feature of the abduction phenomenon? What were they doing?

The Breeding Program is Recognised

Gynaecological examinations have been reported as part of the table procedures so frequently that they have come to be seen as universal. Female abductees often reported a thin needle-like device being inserted into the lower abdomen about the area of the ovaries. After the development of laparoscopic (minimally invasive) surgery in human medicine, it was suggested that these early needle-into-the-abdomen insertions—like those reported by Betty Hill in 1961—were probably a kind of alien laparoscopy. However, this laparoscopy theory does not fit the facts.

We became briefly acquainted with Dr John D. Miller in Chapter Four, where he discussed the possibility that the abductors were siting microimplants in the area of the olfactory nerve where it passes through the mucous membrane right at the top of the nasal passages. Miller posits that this nosebleed-inducing procedure may allow the entities to monitor certain brain activity of their subjects—post-release, as it were—while being very difficult to access, remove, or even detect through normal patient examination or even surgical intervention.

Dr Miller presented his findings in 1992 at the Conference on Abductions held at Massachusetts Institute of Technology (MIT) after investigating firsthand reports of the medical procedures reported to be carried out by the 'alien doctors' and how they differ from those of normal human medicine.

> As a physician, the most consistent impression I get from accounts of alleged alien examination techniques and "medical" procedures, is that I'm not hearing about "our kind of medicine" ... the most consistent feature in the great majority of alien medical techniques are either grossly or subtly different ... bizarre differences predominate over vague similarities.

The differences between reported alien and known human medical techniques and procedures are great enough to invalidate any theory that these reports somehow originate in the witnesses' own past medical experience or knowledge. The "exam" and certain repeatedly reported medical procedures are seemingly the central feature to most abduction experiences and yet to me this central feature is not a derivative of human experience.

The most consistent explanation the alien doctors give for their actions seems to be something along the lines of the mere assertion that "We have the right to do this".[7]

Specifically addressing the issue of whether a laparoscopic surgical procedure explains the reported needle-inserted-into-the-navel procedures frequently carried out by the abductors, Dr Miller makes the following observations:

> Betty Hill reported the insertion of a thin needle-like device into the area of her umbilicus (Ballard 1987: C–81). Since then many other witnesses have reported similar experiences. After the human laparoscope was invented some years later it was widely speculated that this represents some form of laparoscopy.
>
> But there are important differences. Today, the human laparoscope is much greater in diameter than a mere needle. It requires a small incision to insert. To perform pelvic surgery a surgeon must insert a second scope or trocar via a second incision [commonly a third instrument is now deployed in laparoscopic surgery through a third incision]. Also to perform surgery the abdominal cavity must be distended by the insufflation of CO_2 gas to permit visualisation of structures and free movement of instruments. I do not hear of gas insufflation events during these alleged "needle-in-the-navel" procedures.[8]

So then, it looks like the alien doctors are *not* performing laparoscopies.

What they were, and are, apparently doing is extracting mature ova through the thin needle. I have personally known three female abductees with medically diagnosed polycystic ovaries—one of these was my mother—and would hypothesise that this medical phenomenon will be either predominant with abductees, or at least very commonly found in this population. I know of no one who has attempted to explore this hypothesis, and it is difficult to see how it might be definitively tested as at least 95% of abductees, and maybe as many as 99%, are unaware that they are abductees. But there we have it: Aspin's First Hypothesis of the ubiquity of polycystic ovaries in female abductees.

Female abductees have sometimes reported pregnancies which could not be explained, as they were either not sexually active or were taking effective contraception. Budd Hopkins had a case of a 13-year-old who was diagnosed as pregnant by her gynaecologist. Her abdomen began to distend and her hormone levels were appropriate for a normal pregnancy. The girl's parents were furious as she was barely through puberty, and she herself was still a virgin with an intact hymen who knew almost nothing about sex and had never had a boyfriend. In the circumstances the parents demanded a termination and an abortion was performed, so the evidence of any possible exotic or unusual characteristics of the foetus was lost. The girl turned out to be a lifelong abductee. (As explored above, there is no other kind except in very rare circumstances, as with Barney Hill or Chuck Rak in the Allagash case.) It was as an adult that she sought out Budd Hopkins, citing multiple abduction experiences.

Sperm has been reported to be extracted from mature males *every single time they are abducted*. Very often, men are reportedly abducted for this sole purpose and are told so by their captors. (It has happened to me.) One would think that, with the millions of sperm produced by the average human male, an advanced alien race would just freeze the sample and use the stored sperm for years. But they repeatedly go back to the well. They even have a medical technique to effectively obtain sperm from men who have had vasectomies (more detail on this later) and have occasionally been known to extract sperm through the use of a kind of electroejaculation stimulation (EES).[9]

We'll come back to this peculiarity in Chapter Eight, to look at the issue from a new angle.

It has been speculated that female abductees were being implanted with embryos created by a kind of IVF. An ovum from the female abductee is fertilised by sperm extracted from a male abductee, to which a third element is added: a small quantity of alien DNA, bio-engineered into the embryo *in vitro*. Or it may be that alien DNA is bio-engineered in some way into the sperm itself, or into the ovum, prior to *in vitro* fertilisation. It is of course impossible to know this precisely, as the aliens do not instruct abductees in the biological complexities of the process. But the results are obvious: the appearance of the early hybrids were widely reported to be a visible mix of human and alien characteristics.

It's at this point you may experience your own 'Felix Frankfurter moment'. But, incredible and outlandish as it may seem, this is what appears to have been happening to us.

By the 1980s, abductees were reporting 'missing' pregnancies. These women insisted this did not have the character of a spontaneous miscarriage, which is not especially uncommon during the first trimester. After between eight and ten weeks, they insisted, they suddenly were no longer pregnant. There was no bleeding, hormone levels would suddenly return to normal, and repeated tests showed no pregnancy. The foetus was simply missing. What were they to make of that?

In Vitro Gestation, and Nursing the Infants

Also during the 1980s, abductees began to report the strangest things. (Yes, I know …) During an abduction event, while 'onboard', abductees began to report being shown rows and rows of containers, in each of which was what looked like an embryo floating in a light brown liquid. Whole enclosures—'rooms'—aboard the craft were allegedly given over to these vessels. Abductees reported that the aliens seemed pleased with the apparent success of this project and moreover, they expected abductees to be pleased about it too, but seemed disappointed and uncomprehending when the abductees recoiled in horror at what they were doing. During a summer 1976 incident I, too, was presented with this apparent satisfaction at their achievements in successfully completing this bizarre bio-engineering project. They were also delighted with, and appreciative of, my unwitting, unintended input to the result: the new babies.

Simultaneously, female abductees, and sometimes males as well,[10] were asked to "hold" and "nurture" a small baby or toddler to his/her skin, to "interact" with it silently for several minutes before handing it back to (usually) one of the 'female' taller grey beings. This strange ritual was reported by some to be the apparent centrepiece of every single abduction event. Sometimes the abductee would be told (telepathically, as always) that this baby was "her" child, and she should hold and nurture it while onboard. Sometimes the abductee would be told to "feed the baby" with breast milk and would discover that she was, indeed, lactating.

Occasionally abductees were told that these babies would not survive on Earth, or anywhere outside the specific ecology created by the aliens for their gestation, though no further information about this was forthcoming from the aliens. These babies (and toddlers) looked like a genetic cross between their human progenitors and the grey aliens. That our visitors might be engaged in a covert program to breed a hybrid race, however improbable, seemed a real possibility.

This breeding element of the abduction phenomenon was first brought to widespread public attention by Budd Hopkins when he published his 1987 book, *Intruders,* which focussed on the Debbie Jordan case. Debbie, whom we met in Chapter Two, was allegedly told by her abductors in 1983 that she could *name* her hybridised children! Other women have reported similar encounters. The later chapters of the book are full of case histories of *other women, other men* whose reported experiences support and reinforce the revelations in the Jordan case. Prior to this, reports of the reproductive focus of the abductions were fairly widespread though no one had investigated the commonly reported themes with sufficient rigour to make sense of it all.

To stretch credulity even further, women who had undergone radical total hysterectomies started to report feeling pregnant, complete with hormonal changes and positive pregnancy tests, which is a biological impossibility. They began to recall the placement (by the taller alien 'doctor') of extrauterine gestational units in the abdomen, describing the thing as "about the size of a grapefruit," where the uterus used to be. After eight to ten weeks, another abduction would follow and the sac and its contents were removed. Even if the abductee did not at first remember this abduction event at the time, she woke up in the morning and "just knew" she was no longer pregnant, soon to be confirmed by a test and hormone levels declining to normal.

Abductees reporting these events are angry and disgusted by what the aliens are doing, and how they are being used. Many questions arise from this. Why don't they rear the embryos *in vitro,* if their technology will accommodate that? Why take the huge security risk of these pregnancies being revealed and the alien agenda exposed, if there is an easier way? There are possible metaphysical answers to this question which we shall explore in a Chapter Nine. How, too, is the foetus sustained in the absence of a placenta or, presumably, an umbilicus? What is the ultimate purpose of this program, and is it possible to estimate its scale and reach? We shall seek some answers to these questions in later chapters.

The Plan Behind the Hybrid Breeding Program?

Abductees reported increasingly regular interaction with groups of more and more human-looking hybrid children. My personal memories include such interactions, from the 1980s to 1990s, of playing with between six and twenty very small children simultaneously and teaching them 'tumbling games.' I have read nowhere any identical account about this (it's possible I might have missed it, because the literature on this subject now is vast), but this is what I vividly remember. I used to think they were just very odd dreams but could not fathom from where in my psyche such dreams might originate. And they were 'dreams' full of very physical sensations like floating, flying, and falling, occasionally waking up with fresh bruises on the body.

The infants became children, which in turn became adolescents. Soon enough, those adolescents had become adults. These children appeared to grow to maturity over a normal human timeframe. Would they enjoy a normal human lifespan as well?

This is where it gets really strange and, again, stretches credulity. Indeed, the idea invites a personal 'Frankfurter reaction'.

The young adult hybrids were further being reported, through the 1990s, to be assisting the abductors. These so-called 'middle-stage hybrids', who have more visibly alien characteristics and would not pass as human in our society, were claimed to be managing the young hybrid children, while the more human-looking ones assisted with the human abductions.

The literature on the phenomenon of hybrids in human society is vast, though anecdotal. They look and sound just like regular people and, it is reported, they can exercise some neurologic control over humans. This ability to control humans does not appear to be as strongly developed as that routinely displayed by the grey aliens, but nevertheless these human-looking hybrids do have these capabilities in sufficient measure to effectively control our thoughts and behaviour.

The hairs found on the floor of my living room in 2015 (see Chapter Four) were almost certainly from a wig worn by a female, middle-stage hybrid being. I remember her coming into the room, very late in the evening, through the garden doors: that is, the garden doors did not open; she came *through* the doors, as if the glass was not even there.

Ingo Swann was a remote viewer who had worked with both Stanford University's psi program and later for the US Intelligence services. He claimed to have once interacted with a human-looking, female hybrid in a California supermarket.[11]

Timothy Good is a professional orchestral violinist who has also been a long-time researcher of the UFO subject in all its complexity and is the author of eight books on the subject. About ten years ago he told me of two interesting encounters that he had experienced. In each case, Good had sent out a persistent mental message, asking whether anyone in the vicinity "is not from here," and could they please make themselves known to him?

In the first instance, Good was sitting in a quiet hotel lobby in Manhattan. He sent out the mental message, concentrating and repeating it. After around ten minutes a lean, middle-aged man with a bronzed Caucasian complexion and wearing an immaculate dark business suit walked into the lobby, sat down next to him, picked up a newspaper and began to peruse its pages. Good sent out the mental message that, if he really was "not from here," could he please put his hand up to his face? Immediately the man complied, very deliberately extending his index finger up the length of one cheek and retaining it in place there. After several minutes of silence, the man returned his hand to the newspaper. Another few minutes passed, and with no further action, requests or communication from Good, the man got up and, without a word, left the building and disappeared into the street outside.

I asked Timothy why he had not engaged the man in conversation. Timothy is a very polite, well-spoken and cultured man with a moral code of how social interactions ought to be conducted and would be disinclined to engage a stranger in small talk. He replied—it's not difficult to understand this—that the unexpectedness and strangeness of the moment, the fact that no word was spoken and no introduction was made, shocked him into a kind of embarrassed bewilderment. The arrival of the stranger, and his manner of arrival, was so unexpected. It was just possible that this was just some guy who walked in off the street, sat right next to him in an otherwise empty hotel lobby, and several minutes later, walked out. What do you say to the guy? How do you initiate a conversation? To say to a total stranger, "So, you're an alien? Where are you from?" would simply not be the thing to do.

In the second case, Good was touring the United States with a London orchestra. The coach in which the members of the orchestra were travelling stopped at a roadside diner for lunch. Similar situation: out went the same message. After a few minutes, a petite and immaculately dressed young woman, in a wide-hemmed black dress and black shoes, walked up to the table where Good and his party were sitting, looked him straight in the eyes, and ostentatiously low-curtsied in front of him. She then straightened up, again looked him in the eyes, turned away and walked off, staring straight ahead. Not a word had been spoken.

These encounters do not on their own prove anything. But they are suggestive of a cohort of 'people' who are (presumably) integrated into human society and who are receptive to telepathic messaging. And, indeed, who may respond, if enigmatically, if one chooses to send out a silent invitation.

The agenda seems to include the integration of hybrids into human society. They need to learn everything about living among us, including how to speak: not learn a specific language, but *how to actually speak, with their vocal cords*, which does not come to them naturally, as they communicate telepathically in their own environment. They need to learn how to blend in and not stand out as too odd, whether in behaviour or appearance.

They cannot do this without the assistance, willingly or otherwise, of abductees. David Jacobs' 2015 book, *Walking Among Us*, is specifically focused on this part of the phenomenon, as the majority of abductees' interactions are now reported to be with these hybrids. (Or 'hubrids' as Dr David Jacobs names those specifically genetically engineered to fit into human societies.) Should you be curious enough to read *Walking Among Us* (for context, it may be better to read Jacobs' two previous published books on the abduction phenomenon first[12]), you may work out whose reported experiences are mine, as all the abductees in the book are identified by pseudonyms. No clues!

The seamless blending into human society of these 'hubrids' appears to be the next stage of whatever this project of theirs is all about. Reports from abductees over the years reveal that the aliens appear to be most delighted if their subjects are unable to tell the difference between the hybridised creations and normal humans. When abductees are shown a collection of normal humans and 'hubrids' together, the most common question asked is: "Can you tell the difference between us and you?" It seems absurd that the most cherished expressed goal should be that their creations pass in normal human society without being noticed as in any way unusual. And yet more absurd still if abductees ought to be excited and delighted when this is achieved, like it was some sort of parlour game.

In their version of an ideal future, the aliens say that "everyone will know his place," and that the future will be "wonderful." And, moreover, they expect the abductees to share their enthusiasm for this wondrous vision of an alien society, stratified and ordered, with everyone content in their allocated role.

Many 'hubrids' preparing for integration frequently express anxiety about the propensity for humans to be unexpectedly violent to each other. They view their personal safety and the crucial importance of their behaviour in not provoking such violence as a prime concern. This personal security issue is the main reason, it is believed, for their reportedly living in groups of (typically) three to five individuals: the better to control any troublesome humans should things get out of control.

There is consistent testimonial evidence that completely human-looking alien-human hybrids have largely taken over the abduction program and are managing the tasks and responsibilities previously carried out by the greys. Some abductees claim that the greys no longer deal with them at all. However, the majority of activity characterising interactions between the newcomers and abductees involves the teaching of skills to ease the hubrids' integration into human societies around the world. They have to learn to look right, dress right, act right, and generally not stand out as odd. As early as 2009, a German abductee of my acquaintance then living in London reported to me that *all* her interactions by that time involved only the hubrids and their supervising hybrid "security minders." (She referred to the beings with whom she interacted not as 'hybrids' or 'hubrids' because she had no familiarity with the literature, but as "the human-looking ones.")

We shall return briefly to this subject in Chapter Nine.

What Does the Government Know?

Elected politicians come and go. After they are gone from office, many busy themselves with their memoirs, sharing all manner of anecdotes and gossip. However, there will almost always be certain information to which they'd been privy but which must nevertheless remain unsaid, under severe penalty.[13] This is necessary so that temporary leaders, ostensibly chosen from among regular citizens, do not go on to reveal state secrets of, for example, intelligence operations or weapons programs and the like. In the UK, a '50-year rule' obtains, prior to which many public records must not be released. Even after fifty years, in fact, many records may remain undisclosed if judged to be "in the interest of national security." Many nation states have Freedom of Information legislation on statute but exceptions to release are always permitted in a variety of circumstances.[14]

A leading investigative journalist (I'll call him Nathan) with whom I am personally acquainted once declared to me that the majority of elected politicians "are unremarkable people of limited abilities." He was referring to members of the UK House of Commons, but the elected politicians of any democracy could surely be described in similar terms.

Perhaps the question, "What does the Government know?" more specifically ought to be: "Is there anyone in a permanent career position within government—as opposed to an elected representative who may be removed at the next election—who might understand what is going on with this phenomenon?" For example, among the permanent high- or upper-middle-ranking members of the Intelligence Community.

While it is impossible to know for sure, I personally suspect the answer to that is "no." Almost nobody who has not become involved with this in some manner knows what is really going on, and *they couldn't possibly know*. The only people in the population who know about the program are abductees, and at least 95% and possibly 99% of them do not understand what has been happening to them and remain in the dark. You need diligence and determination to find out about all this, while avoiding all the nonsense and deception surrounding the subject. Most people do not know where to begin and lack the courage and fortitude necessary to make any real progress with it; those who persist and struggle through into the sunlight discover that incredulity and dismissal (and sometimes worse) greet any public admission of their experiences, because "it just can't happen."

We'll return to examine some of these issues from a deeper perspective in Chapter Eight, but for the moment the following accounts might offer some small clues that "someone, somewhere" seems to think this subject might be important and is paying attention.

Someone Taking an Interest in My Taking an Interest

What follows are some of my experiences of nuisance intervention when my personal interest in the phenomenon has come to somebody's attention.

Snooping My Mail

In late 2007, I ordered a copy of *Alien Discussions*,[15] a massive, 683-page tome which covers the academic findings presented and discussed at MIT during a conference on alien abductions in June of 1992. It's extremely hard to find, and the copy that I had tracked down was for sale by a bookseller in Germany. It was expensive and rare.

The transaction was a private sale between myself and a German bookstore, a buyer who'd deliberately avoided any public exposure and had kept a very low profile.

The book should have taken no more than three or four days to arrive. When it finally arrived six weeks later, it had been repackaged in a polythene bag and covered in yellow Customs and Excise tape, with a label announcing that the UK customs authorities had intercepted the item as suspicious. The delay was the big issue. Although the book had arrived undamaged, it had been held for *five extra weeks*, which is well beyond the time that a package from Germany ought to have taken. What the **** was that about? And on whose orders was this delivery delayed for so long?

The year following, in May 2008, I ordered a copy of *Flying Saucers and Science* directly from the author, the late Stanton Friedman.[16] Stanton lived in Canada, from where he'd shipped a signed copy to me in the UK.

Same ritual as with the MIT book: five-week delay, book arrived opened by UK customs, re-packaged in a polythene bag covered in bright yellow Customs and Excise tape, and with the same pattern of label as before, bearing the same message.

Over the years I have purchased and imported hundreds of items, and only these two books have ever been intercepted. The first was an expensive and rare academic book on this topic, which would be sought out and purchased only by a serious researcher into the subject. The second was sent directly to me by the author, a high-profile researcher in the field who gave regular public lectures to university audiences. This looks like two examples of minor harassment and indicates that someone wants to inform me that they've 'got my number'. Why the agency responsible resorts to this ridiculous behaviour and doesn't just come around to meet me, is difficult to understand.

The Very Real, Years-Long Surveillance of Robert Hastings

Some of the most blatant examples of low-level harassment have occurred during the time Robert Hastings and I have been together or have attempted to communicate. Robert has detailed numerous examples of 'someone taking an interest' in *his* taking an interest in the connection between UFOs and nuclear weapons.

As mentioned earlier, Robert had accompanied my wife and me on a trip to Normandy in September 2014. When we returned to London, he had arranged to meet up with 'Nathan', the investigative journalist referred to above. He had demonstrated a continued interest in Robert's work investigating and cataloguing UFO incidents at nuclear weapons sites, among them the apparent interference with intercontinental ballistic missiles.

We arranged to meet on Monday 15th September 2014 in a pub in Richmond-on-Thames, as Nathan at that time resided with his family in southwest London. The liaison was arranged via email and on the Monday evening, off we went to Richmond.

As the four of us settled in for a discussion at a corner table, a lean, fit-looking middle-aged man came in and sat down at a nearby small table. He was about two metres away from us, and as it was a Monday evening the room was not crowded. The man rather ostentatiously took out an electronic box about the size of a vintage cassette player from the 1980s, put on a large pair of headphones and settled down to stare at us. After about 30 minutes, the man got up and left, but was immediately replaced by another, who behaved in the same manner.

Nathan and Robert had a long and detailed discussion for more than two hours, after which the four of us stood up, and left together. The remaining stranger immediately removed his headphones, packed up his gear in a shoulder bag, and followed us out. The first man appeared, walked very close behind us and (seemingly intentionally) stood on Robert's shoe from behind, causing it to spring off his foot. Not only did he not apologise but he continued to shadow us in silence until we'd reached our respective parked vehicles, where we said "goodnight," and went our separate ways.

Robert's account of this incident is more comprehensive than my own recollection:

> Hi Steve,
> ... what I recall is that as we all—including [Nathan]—were leaving the pub, and climbing up a long flight of outdoor steps, that led from river-level to street level, a guy walking behind me stepped on the back of my shoe, forcing it off my foot, requiring me to stop momentarily to pull it back up. I recall looking at the guy and wondering why he was walking so closely behind me. At the top of the steps, we said farewell to [Nathan] and he walked away in the opposite direction. I don't recall where the other guy went but he wasn't following us anymore.

At that point, Jan said to me, "Did you see the two guys who were sitting right behind us, at the next table? They seemed to be listening to you and [Nathan] very closely." I had not been aware of them so I asked her to elaborate. Apparently, one guy came and sat down just after we ourselves were seated. At some point, Jan became aware of his odd behavior—his leaning in, just behind my back, as if he was trying to hear what [Nathan] and I were saying—and began staring at him. Eventually, the guy noticed Jan looking directly at him and quickly got up and left. However, she said, another guy quickly came over and took his seat—the exact same seat—and also seemed to be listening to us. I recall that the room was mostly empty and many other tables were available. But, for whatever reason, these two guys wanted to sit next to us.

Jan said that it was the first guy who was walking close behind us as we left the pub. I was stunned by all of this info and asked her why she hadn't brought all of this to my attention while it was happening. I don't recall her response.

In June 2013, I had written an article [17] in which I proposed that the supposedly comprehensive release by the MOD of its UFO files was probably a smokescreen and actually got Nick Pope to add a few comments that generally supported my contention. Given that my email is clearly, probably continuously monitored by, I presume, the NSA or CIA (or perhaps AFOSI), I have concluded that one of those agencies shared my article—and my emails to/from [Nathan], setting up our meeting—with the MOD or MI6, or some such group, and it was that particular British organization that tasked the two mysterious guys with monitoring my conversation with [Nathan]. Perhaps they thought that he was going to write an article about my work, or my claims about a UK government cover-up on UFOs. Who knows?

Robert

Since that time, Robert and I have had several conversations via Facetime. All is OK for a few minutes after the initiation of the call, but then the line drops. We try to reconnect, and whether he is calling me, or I him, we each receive a message that there is no service. As an experiment, we have tried calling a third party, such as a family member, and the call will link up without problem. Service is interrupted only when we attempt to speak with each other.

Robert details several such incidences of surveillance in his own book on this subject, *Confession*.[18] There have been several consistent email

Robert Hastings with the author in London, 2014.

'anomalies'. Messages between several specific individuals and me are frequently found in the junk mail folder, or discovered in the 'deleted' file, or else filed with a date two years earlier so can't be located or disappear without trace.

The correspondents are, exclusively, individuals with a public profile concerned with the alien abduction phenomenon; these anomalies never occur when attempting to communicate with anyone else. I first noticed several years ago when corresponding with Dr. David Jacobs at his official academic address at Temple University, I receive no response. After repeat-sending the same message I would call him; he would on investigation discover that all my messages had been confined to the 'junk mail' folder on his computer. He claimed to never see this anomaly with other correspondents. Same phenomenon with Robert Hastings, though due to the regularity of our interactions – often several times each day over the past 12 years through a variety of media channels – this intervention is applied on a more selective basis, dependent on the substance of our exchanges. It took Whitley Strieber and me several months to arrange an online interview, so frequently did the messages 'disappear'; in the end Whitley arranged a meeting through an intermediary known to us both, someone whose interest in the abduction phenomenon is less prominent and publicly known.

It is difficult to make sense of any of this. What is certain, however, is that this interference with communications is real and consistent. The 'Why?' question opens a box of endless speculation. The conclusion must be that someone, somewhere, is telling us 'We are aware of you and your interest in this issue; it's important enough to make you aware that we can hack your phone and email, intercept and read your messages'.

The first edition of this book was published by KDP on 12 September 2022. It sold well for six months, with monthly royalties paid retrospectively after 60 days. On 31 March 2023 my KDP account was suddenly closed down with no warning, with author access denied. The 'explanation' offered by KDP was spurious and provably untrue. At the same time, a number of other authors specialising in this field of study also had their KDP accounts closed and were offered the same spurious 'reasons', with all their books removed from sale. All appeals were rejected, none of the accounts were restored and all royalty payments cancelled. I have never been attracted by conspiracy theories and am generally contemptuous of 'believers' in such, but something very strange is going on here that makes no sense.

Are UFO Incursions at Nuclear Weapons Sites Related to Abductions?

Several USAF and other personnel who have gone on the record regarding UFO incidents at nuclear weapons sites, some of whom testified on the subject at the National Press Club news conference on 27th September 2010 in Washington DC,[19] have themselves been abductees, though few have gone public about it, choosing to discuss it only in private.

What we can say with some confidence is that, in the light of the abduction program, if the plan of those responsible for the Program is to stay here on Earth long-term, they are likely to wish that the human race does not annihilate itself, despoiling the planet in the process. Judging by the examples of numerous reports of UFO incursions at nuclear weapons launch and storage facilities during which missiles have been shut down or otherwise interfered with, it appears that this is in fact well within their power to do.[20]

In the next chapter we examine the life and history of Budd Hopkins, the great pioneer of research into the abduction issue. His brilliant, astute intellect could see directly into the heart of the subject, uncovering much that had previously been unknown—and unexpected—prior to his engagement with the issue. During his final years with us he taught me much about the phenomenon and how it works. Moreover, he proved in person to be a compassionate humanitarian and an exceptionally brilliant man with a sharp mind and fine sensibilities, who could engage knowledgeably on a wide range of subjects unrelated to the study of the UFO and abduction phenomena.

Chapter Six

The Great Pioneer

Personal time spent with Budd Hopkins during his final years

Most readers interested in, or even just mildly curious about, the subject matter of this book will have heard of Budd Hopkins. Even if you haven't read any of his five published books, nor watched any of his public lectures, nor seen nor heard any of the hundreds of interviews he gave over thirty years on television, radio, and online, you will probably know of him. Indeed, he will be forever associated in the public mind with his ground-breaking studies of the alien abduction phenomenon. He was a pioneer of the research, and an articulate spokesman for all of the investigators who came after him—not to mention to the thousands of people around the world who've found themselves at the centre of this mystery.

The irony is that Hopkins' primary and very successful career was that of a professional artist. He came to the UFO/abduction subject only in mid-life, initially out of curiosity, following his own daylight sighting of a UFO.

This chapter tells the story of my blossoming friendship with Budd Hopkins during the last two years and eight months of his life, before his death in August 2011, aged eighty. We first became closely acquainted in the final months of 2008. During those years I stayed with him in NYC on several occasions, and we spent many hours on the phone together during his final days with us.

Brief Bio

Elliot Budd Hopkins was born in June 1931 in Wheeling, West Virginia as the middle child of three (an older brother Stewart; a younger sister Ellie) to a conservative suburban family. As Elliot was also the name of his father, Budd was always known only by his middle name.

In discussion with Budd at 246 W16th, May 2010.

He described the social ecology in the provincial WV town into which he was born and spent his formative years, thus:

"No African-American man, woman or child, other than a menial employee or servant, ever crossed our threshold. The country club to which my parents belonged was 'restricted', a euphemism meaning No Jews Allowed, and almost no-one I was to know for the next ten years of my life felt anything but hatred for President [Franklin] and Mrs. [Eleanor] Roosevelt. In short, I was born into the antediluvian Middle West." [1]

To further illustrate the prevailing social and political landscape in the WV of his childhood, Budd once recounted to me the story of an event which took place when he was nine years old. The November 1940 federal elections were under way. FDR had already served two successful consecutive four-year terms between 1932 and 1940 and was running for a third. Henry Wallace was his chosen VP. Budd's father, with country club friends, declared themselves to be "poll watchers," and stationed themselves outside of the voting stations in Wheeling for the whole long Tuesday of the November 1940 election. When the young Budd asked why they were spending the day doing this, his father told him bluntly—and with commendable honesty—"To stop the niggers from voting." Budd never forgot this.²

The infant Budd contracted poliomyelitis at age two, and spent a year with his right leg in callipers; unsurprisingly, he became a keen advocate for vaccinations when the first successful polio vaccine finally became available during the 1950s. That infant polio infection left him with a slight limp. Although he could walk unaided for quite long distances and stand in front of an audience when delivering a two-hour lecture with little apparent discomfort, this minor disability was sufficient to fail the US Army medical, excusing him from military service when drafted for the Korean War in the early 1950s. Unable to run fast, as a young man he found engagement in active sports difficult. The exception was swimming, as the water rather than your legs supports your body weight, and at swimming he excelled. Many hours were spent each summer swimming in the Atlantic off Cape Cod, where he bought a summer home to build a second studio in Wellfleet in the early 1960s, and would spend the last fifty summers of his life.

During his year of enforced polio confinement, the young Budd learned to draw and paint and an early natural talent for visual art was discovered. His schoolteacher told him that he "should be an artist one day." Although he immersed himself in painting and drawing, "No one told me in those early years about art with a capital 'A,'" and it was not until his 19th year, when at Oberlin College OH, that he "for the first time heard the names Van Gogh, Cezanne and Picasso." ³

During these Oberlin years Budd's eyes were opened not only to the world of fine art but, in the multicultural milieu of a liberal Ohio town in the 1950s, to liberal-progressive political ideas. His impassioned commitment to both never faltered for the rest of his life.

After graduating from Oberlin in 1953 Budd moved to New York City, attracted to its vibrant School of Abstract Expressionists. Robert Motherwell, Willem de Kooning, Franz Klein, Joan Mitchell, Mark Rothko, and Jackson Pollock were among his contemporaries and friends. These pioneers of art in mid-century America effectively moved the creative focus from Paris to New York. After a few years of struggle, Hopkins eventually secured his first exhibition and soon the young artist began to make a living from his talent.

In 1976, Hopkins was awarded a Guggenheim Fellowship and shortly after that a substantial grant from the National Endowment for the Arts. He wrote and published articles in art magazines and lectured widely on art history in schools, proving to be a natural public speaker with engaging insight and humour. In 1994, he was invited to become a full academician at the National Academy of Design, an offer he gratefully accepted.

He was still creating ground-breaking art in his late 70s. Notable Hopkins collections are at MOMA, the Met and the Guggenheim, all in NYC. Prominent private collectors of Hopkins' work included the actors Omar Sharif and Maximillian Schell, both of whom had been personal friends. When you got to know Budd well, he revealed a treasure trove of affectionately hilarious stories about his dealings with such people and proved to be a fine raconteur. Budd told the story of hosting a small dinner party in late 1969 for Max Schell to introduce him to Mark Rothko, whose work he greatly admired, shortly before Rothko's tragic suicide a few short months later.[4]

So much for Budd Hopkins' background, how he arrived in New York City, and the nascent but highly promising artistic career engaging him prior to his interest in the UFO subject being awakened.

The Journey to Abduction Research

There is nothing like a clear, daylight sighting of a 'flying saucer' in the sky above, especially if witnessed in the presence of others of sound mind, to pique one's curiosity about the phenomenon (my incurious father perhaps excepted). For Budd it was a clear daylight sighting of a classic lens-shaped UFO in the sky over Cape Cod early one evening in August of 1964.

He was in the company of his first wife—Joan Rich, who was married to Budd for thirteen years—and a visiting friend from England, Ted Rothon. At the urging of his father-—with whom he admitted to an often fractious relationship due to their political differences—he belatedly reported the sighting to nearby Otis Air Force Base. In response, he received the intriguing request that, should he have another sighting, would he please report it immediately, without delay, "to the nearest Air Force Installation." [5]

Following the 1964 sighting, Budd recounts:

> Over the decades, I can see that virtually everything I painted for the next twenty years contained a large, dominating circle of some sort: black, colored, divided into pie-slice sections, banded, repeated in fragmentary form, and varied in every way I could think of ... one critic remarked that

I "seemed to be painting portraits of circles", a canny observation in that I usually located the circle above the mid-point of the canvas and to the left or right, in exactly the kind of off-center area where a portraitist might place a sitter's head. Like the infinitely varied human face, I wanted my geometrically regular 'circle portraits' to seem almost as different from one another.[6]

The first in this long series of post-sighting circle portraits, titled *Sun Black*, was completed in 1966.[7]

After a decade of interest in the phenomenon reading books, newspaper and magazine articles, and examining published photos of UFOs and so forth, in 1975 Hopkins carried out an amateur field investigation himself, assisting the experienced investigator Ted Bloecher.[8] This was the case of a UFO landing, during which around a dozen small humanoid occupants exited their craft, in North Hudson Park, North Bergen NJ—directly across from Manhattan—in January 1975. The eyewitness was George O'Barski, who owned a liquor store across the street from Budd's home on w16th Street. Budd described O'Barski as "a grouchy old guy, who had seen everything in life and resented most of it," and who was emphatically not the kind of person given to fantasies or invention.[9]

Several additional witnesses were tracked down and interviewed, including three night doormen from the nearby *Stonehenge Apartments* building who had seen unusual lights in and around the park on separate nights. During O'Barski's encounter, a doorman was watching from the lobby when a large pane of glass was cracked, coinciding with a high-pitched whining, shortly before the UFO abruptly departed at around 3am. O'Barski took the investigators to the location where he had observed the UFO land and the diminutive occupants emerge to dig holes in the ground, apparently to collect soil samples. O'Barski had returned the next day to find the holes precisely where he'd seen this occurring. Months later, the ground still showed clear signs of the holes, though they'd been filled in by groundskeepers, when the investigators were brought to the scene.

Budd submitted the results of this long and thorough field investigation as an article to *The Village Voice* in early 1976. *Cosmopolitan* magazine bought and reprinted it a few months later, "sandwiched in between one article on how to have an orgasm and another on what to wear on your first date," was how Budd liked to tell it with his typically dry, risqué humour.

One of Hopkins' large 'Sun Black' portraits, right, photographed in his Manhattan studio in September 2011.

The case became well known and excited a great deal of media interest. Hopkins remembers:

> The cumulative effect of all this on my life was inescapable. In response to the general public interest in this important, well documented case, I began doing interviews and speaking out on the reality of the UFO phenomenon. ... When I walked into a gallery opening shortly after my 'Voice' article appeared, several friends came over, eyes wide, and told me they had seen me on the local TV news the night before. TV was not the place anyone expected to see a painter in 1976, unless he had committed some kind of heinous crime and was doing the perp walk in an orange jumpsuit. Naturally my friends at the opening wanted to hear more about both the landing in North Hudson Park and the UFO phenomenon in general. "Is there really something to it?" they asked ...[10]

Tens of thousands of people each year all over the world experience sightings of and interactions with UFOs, but there is generally nowhere 'official' for them to report or discuss these experiences. Before the advent of the internet, when an article on the subject, whether in general or focusing on a particular case or incident, was published in a print journal the likes of *Cosmopolitan*, with the wide circulation and huge market presence it enjoyed in the 1970s, it inevitably uncovered cases hitherto not widely shared. Witnesses who had not had the opportunity to share their experience with some authority of the subject will naturally contact the author of the piece because s/he assumes said author to be the 'expert' they'd longed to speak with. At the least, they will feel confident of a sympathetic ear. (These days the worldwide web has modified the way this works, but there is still nowhere 'official' to report to and if anything, the landscape in the internet age is even more confusing).

During the months following *Cosmo*'s publication of the article describing the Hudson Park NJ incident, hundreds of people wrote to Budd Hopkins about their own UFO sightings. The meticulously researched article had ensured the author was now regarded as a bona fide UFO investigator—whatever that counted for. (Almost everyone doing such investigations was an amateur enthusiast of one kind or another.) You may read the history of this case and of the resulting torrent of mail in the first chapters of *Missing Time* and again, from a slightly different perspective, in Hopkins' memoir, *Art, Life and UFOs*.

Although an artist, Budd brought a genuinely sceptical, enquiring scientific mind to the UFO subject. He was astonished by the number of correspondents who wrote to him about UFO sightings concurrent with a period of 'missing time' and who remained disturbed by the feeling that "something strange happened" yet could not rationalise it. Intrigued, he phoned and subsequently met with many of these witnesses. His high intelligence and brilliant artistic sensibilities enabled him to begin to discern patterns in their narratives.

Hopkins recognised that it was essential to access these memories if at all possible and if the retrieved memories were going to be reliably accurate. Taking his cue from Dr Benjamin Simon, who treated Betty and Barney Hill following their 1961 encounter in New Hampshire, Hopkins personally did no memory retrieval work himself during the first seven years of his investigation into abductions but began to refer these witnesses to qualified psychologists, notably Dr Robert Naiman and Dr Aphrodite Clamar who both had an interest in the subject and offered their services for free. Hopkins was not physically present for either the consultations or the recorded interviews.

Hopkins later traded his own artworks in exchange for the psychologists' time. This is how practising psychologists in NYC got to know of Budd's work with this subject, as many of them were dealing with clients reporting traumatic anomalous experiences which, in hindsight, proved untreatable with conventional psychiatry but closely matched the abduction narrative. The stream of potential abduction cases multiplied; the phone rarely stopped ringing and the mailbox each day was full.

Scores of people wrote to Hopkins, troubled by a missing time episode coincidental with a UFO sighting. He came to strongly believe that the abductors buried memories of the encounters.[11] Although such a manipulative neurological process is currently beyond our medical technology it seemed reasonable, given the evidence, that it wasn't beyond the talents of a race of beings that could travel between the stars. It began to seem to Budd and others that this memory obstruction served to allow their activities to remain covert.

Usually nothing specific is remembered except, perhaps, the recollection of a close-up UFO sighting. An uneasy feeling remained, that something disturbing had happened, perhaps on a stretch of road along which the witness had been driving, and that some hours or so had mysteriously passed unnoticed. Many are left with a very disturbed feeling about the UFO subject. This serves the abductors' purposes very well, ensuring the human population in general continues in ignorance and complacency, taking no action to stop the program.

When *Missing Time* was published in 1981 it immediately revolutionized ufology. Its careful, thorough examination of seven cases, focusing on the common themes between them, brought the abduction subject to a wider public and moved the centre of gravity from mere sighting reports to the question of the potential purpose of the UFO phenomenon. Budd Hopkins had become an acknowledged expert on UFO abductions. He was interviewed on the main TV networks and invited to UFO conferences, where he was invariably among the top attractions. In this role he excelled, as he was an engrossing and entertaining public speaker with years of training in delivering lectures to large audiences at art colleges, universities, and gallery events.

The art took a back seat for a while, even though it continued to be the main source of income for the family: Budd had produced a daughter in 1973 and married April Kingsley, his second wife and the new baby's mother, the same year. He dedicated *Missing Time*, "To April, my Loving, Patient Wife."

Hopkins presented the following common patterns in those reporting abductions, all of which are now universally recognised as essential components of the abduction phenomenon:

1. All the witnesses reported having experienced a period of missing time

2. All had a disturbing feeling deep down that something really strange had happened to them during the missing time period, which persistently bothered and worried them

3. Many of them had bodily scars or marks of distinctive pattern, all in the same areas of their bodies: straight-line scars, scoop marks, sometimes bruising around the ankles and lower legs, sometimes what looked like needle incisions in regular patterns, typically on the forearm or hand in the form of an equilateral or isosceles triangle

4. When abductees reported persistent nose bleeds, Hopkins was the first to note the significance of tiny nasal implants in near-identical sites not easily accessible by conventional surgery when revealed by x-ray

5. Hopkins first postulated that abductions might be more commonly found in the population than simple UFO sightings, of which tens of thousands are reported worldwide each year

6. He first postulated that abductions seemed to be repeated in certain individuals, not random events, over the lifetime of the abductee

7. He was the first to confirm that abductions were intergenerational, and that virtually every abductee had at least one abductee parent, or occasionally both

8. He was the first to establish that the commonly reported sperm and ova harvesting was for the deliberate purpose of the breeding and gestation of hybridised offspring which were partly incubated *in vitro*, for future purposes unknown

9. During his long years of investigation he discovered that some abductees are brought together to (it is assumed) enable the entities to explore and understand human relationships[12]

Various examples of the latter subsequently emerged, such as the so-called 'Mickey and Baby Ann' bondings, as well as same-sex and friendship type bondings as described by Beth Collings and Anna Jamerson in their excellent work *Connections*.[13] The award-winning journalist and investigative reporter C.D.B. Bryan devotes a long section of his book, *Close Encounters of the Fourth Kind*[14] to interviewing 'Carol' and 'Alice', pseudonyms for the co-authors of *Connections*.

Hopkins was extremely patient and methodical in his research, and eschewed the temptation to publish findings based upon a single case, or even just two or three. Multiple examples, rigorously examined from every angle, were always required before publicising new information. He always counselled researchers to "look for patterns and pay little heed to the one-off or outlier experience." He firmly believed that the common patterns offered the key to understanding.

He always worked with others, most notably in the early days with fellow researcher Ted Bloecher and professional psychologists Aphrodite Clamar and Robert Naiman. From 1979, he attracted a small army of volunteers to staff his Intruders Foundation to help respond to the avalanche of correspondence, process donations, and ease the administrative workload. (As I was to later discover, keeping up with correspondence and administration was his Achilles' Heel.) Due to the thoroughness of his ground-breaking research and well-written books on the subject, which revealed and explained the phenomenon in a lively, intelligent and engaging style, he attracted many others into the field including university professors Dr David Jacobs and Dr John Mack, along with scores of others.

Hopkins' four published books on the abduction subject are *Missing Time: A Documented Study of UFO Abductions*; *Intruders: The Incredible Visitations at Copley Woods*; *Witnessed: The True Story of the Brooklyn Bridge UFO Abductions*; and, with Carol Rainey, his third wife, *Sight Unseen: Science, UFO Invisibility and Transgenic Beings*. A fifth book, *Art, Life and UFOs* is a memoir of life experiences—a mini-autobiography not intended to be revelatory or instructive about the abduction phenomenon. It is nevertheless an entertaining and highly instructive read about the man himself, in which his essential character is laid bare. It's also at times delightfully witty and entertaining.

Hopkins also was the author of many academic papers and essays on the subject, most of which were published in periodicals or in conference summaries. He lectured widely to groups of therapists and investigators of the phenomenon and, over twenty-five years, gave more than one hundred conference presentations. Along with his many interviews the archive of his work on the subject is wide-ranging and substantial.

He was a true pioneer and would prove to be, in person, even more of a remarkable, brilliant, warm and compassionate humanitarian than I could ever imagine he would turn out to be. He was unique, a one-off in the best sense. I am very grateful to have known him. But getting to him in 2008 proved to be a challenge, as I shall now relate.

My Long Journey to Meet Budd Hopkins, or at Least to Get His Attention

The following is illustrative of the considerable lengths to which an abductee will go once s/he accepts what is likely going on. There comes a moment when one feels the need to connect with others for whom this mystery means something. No mere shoulder to cry on, or friendly ear to bend, but someone who *understands*. Somebody who has already recognised the phenomenon as being real, and who is actively engaged in trying to figure it out. Earnest explanations to acquaintances feel like a waste of time, however sympathetic the audience. This thing intrudes into your life from infancy and affects, to a greater or lesser degree, everything you do and think, your state of physical and mental health, attitude to relationships, to career, obsessions and deepest thoughts about life. Add to that the *alien* factor and it begs for much more than a solemn heart to heart over a pint at the local.

Once the penny dropped, off I went and there was no stopping me. It was time-consuming and expensive and took me away from other responsibilities, many of which were put on hold, because I realistically had no choice. I had finally stumbled on the cause of all the problems, all the weirdness, all the debilitating health issues. I soon had set my mind toward learning all that I could, and to find someone with whom I could share my experiences. It was quite a journey I can assure you.

I came across the name of Budd Hopkins initially from the pages of Whitley Strieber's *Communion*, which had been published in 1987. Strieber was a successful fiction author with a substantial public profile in the 1980s, with a couple of his books having been made into feature films. Like Budd, he was a Manhattan resident. Hopkins' first book, *Missing Time*, had been published in 1981 and the author was already known to the Manhattan medical psychiatric community and respected as an emerging authority on these weird UFO abductions. Strieber had approached Budd to ask for help investigating his abduction experiences, as subsequently detailed in *Communion*.

At no time in his life did Hopkins ever charge any abductee any money for anything, ever. Strieber, despite his considerable personal wealth due to his success as a writer, like all other abductees who came to Budd for help, was charged nothing.

So, after reading *Communion*, twenty years went by during which my life was taken up by other things, and the subject of alien abduction was rarely at the forefront of my concerns. Then, as explained at length in Chapter One, something suddenly shifted and I was driven to investigate, initially, exactly what might have happened on that early morning of 22 July 1972. This 'trigger event', no matter how much I tried to ignore it or brush it off as "one of those weird things which happened to me" always bothered me deep down but I had for years no idea how to begin to understand it. It is common for abductees to experience one event in particular which they later describe to themselves, and sometimes to others, as a one-off: "When I was sixteen (or twenty-three, or twenty-eight, or whenever it occurred) I think I may have been abducted by aliens."

In a few short weeks in late 2007, I sought out and read all four of Hopkins' then-published books in their chronological publication sequence. Hopkins I discovered to be a highly literate and engaging writer who got to the core of this seemingly outlandish subject with admirable clarity, humanity and frequently a touch of genuine humour. So astounded and impressed was I that someone had engaged effectively with this subject and seemed to understand its essentials so well that I walked round in shock for several weeks, struggling to focus on my professional responsibilities. As managing director of my successful surgical innovations company during this period, I doubtless appeared utterly distracted both to my staff and to the clients with whom I worked.

Determined to meet this man who had clearly come to understand what had been happening to me and had explained it so well, I was filled with a new resolve. It was evidently a pattern common to many, possibly thousands, of others and in no way a rare event as had been assumed when the first early cases gained wide publicity.

Simultaneously I attempted to contact Dr David Jacobs, then a tenured professor in the Department of History at Temple University in Philadelphia. I read his two truly excellent books on the abduction phenomenon. (In 2015, there was to be a third, in which I would be a featured, if anonymous, subject.) Jacobs, too, it turned out, had become deluged with literally thousands of letters from people desperate for help—or even simply acknowledgement of their troubling personal mystery. Whereas Hopkins had eventually been forced to set up his foundation and hire volunteers, Jacobs prepared a detailed questionnaire and rigorous screening process through his International Center for Abduction Research (ICAR) website.[15] I received a brief response from Jacobs in October 2007, though it was limited to the subject of a very specific question and it was a year or more before our conversation would continue.

Hopkins, seventy-six at the time, was by the winter of 2007-08 suffering from lymphoma and undergoing medical treatment. Unaware of his personal circumstances, I wrote to him but received no reply. The email address for his Intruders Foundation similarly produced no response. Then, in February 2008, I travelled from my Hertfordshire home near London to a MUFON conference in Atlantic City NJ where Hopkins was billed as one of the speakers, in the vain hope of meeting him. Unfortunately, he did not attend the conference due to illness.

The transatlantic flight I had taken was from London to Philadelphia, from which on the map seemed a fairly straightforward drive to Atlantic City, and looked less complicated than driving there from any of the NYC airports. I had also hoped that I might fit in a visit to Washington DC, which is closer to Philadelphia than to NYC. I was aware that David Jacobs lived in Philadelphia but at that time did not know him and had no idea where exactly he lived.

The weather in Philadelphia in February 2008 was freezing. Inconveniently, I was detained at the airport and questioned by over-zealous immigration officials about certain scary entry-visa stamps in my UK passport. During the previous six months I had made business trips to both the People's Republic of China and the Islamic Republic of Iran, which exercised the US agents considerably. My bags were thus thoroughly searched prior to my being granted admission to the Land of the Free. I cooled my heels while this pantomime played out (they searched literally *everything*) but after something like two hours was in a rental car on my way to Atlantic City which, unfortunately, was barely any warmer than Philly.

Despite the disappointment of Hopkins cancelling, and the problems with Immigration, the trip wasn't a complete wash. Compensations included having dinner one evening with historian Richard Dolan and his then-wife Karen. (Very good company they turned out to be and I subsequently dined with Richard on further occasions during the following years, including one memorable evening in an Argentine steak restaurant in Leeds.) I also met another of the conference speakers, Peter Robbins, who knew Budd and had worked with him for years through the Intruders Foundation, and we exchanged contact details.

Prior to flying home I decided to spend three nights in nearby Washington DC which was only a few hours' drive away. In DC, I visited the US Capitol building to watch a session of Congress from the viewing gallery. Nancy Pelosi, a several-times re-elected Congresswoman for a San Francisco district, had by then become the first woman Speaker of the House in US history and so third in line of succession to the POTUS, and she presided over the session. Finally I slogged through the falling snow to pay the first of many visits to the fabulous and truly world-class Smithsonian National Air and Space Museum, which had always been on my shortlist of places to see and which exceeded all expectations.

After spending a lovely three nights in Washington, during which I soaked up the energy of that quintessential American tradition, the presidential election—this time, to be ultimately fought out between Barack Obama and John McCain—I delivered the rental car back to the airport in Philadelphia. Before returning to London there was a long delay caused by a brutal snowstorm, the severity of which had temporarily grounded all air traffic. We were kept in the aircraft cabin on the runway with the doors closed for seven hours. All the airline food for the transatlantic flight was served before we ever left the ground. I spent the time crammed into an economy class seat reading Dr John Mack's *Passport to the Cosmos*, which I found to be an 'interesting' take on the abduction phenomenon, but overall less informative and compelling than the seminal works of Hopkins and Jacobs, all of which I had read in the preceding months. Budd's comments on the 'Freud vs Jung' issue from Graeme Greene's memoir, *A Sort of Life,* and how they relate to John Mack and himself are quoted later in this chapter and are particularly pertinent here.

My desire to meet Budd Hopkins continued to be frustrated.

Although I would attend two more conferences in the US during the spring of 2008—in Arkansas and in Gaithersburg Maryland—still no meeting with Hopkins. However, these journeys were not to be altogether wasted.

In Arkansas, my almost unbroken record of rental car travel through the United States in horrendous weather conditions continued. Landing in Fayetteville in the northwestern part of the state, as I sought out the rental car booked for my journey to Eureka Springs in the airport car park, lightning flashed in the fading twilight as the car park's floodlighting came on in the gathering gloom. The rental car turned out to be a huge barge with front bench-seating for three abreast, complete with Texas license plates.

I drove for four hours through the night from Fayetteville to Eureka Springs through a storm with hailstones as large as golf balls crashing onto the car roof with a truly deafening racket, the only vehicle on the highway driven by the only driver apparently foolhardy enough to be out in such a storm. On arriving exhausted in Eureka Springs, the first speaker at the conference whom I met was Timothy Good.[16] Linda Moulton Howe was also there. She was a former investigative TV reporter who had looked into the widespread cattle mutilations during the 1980s, culminating in a disturbing documentary on this issue titled *A Strange Harvest*. I told her about the mutilated cow I had seen in 1970 as a 14-year-old in County Kerry. She was mildly surprised, as she had at that time heard of no reported cases in Ireland. I told her the local vet had diagnosed anthrax, and she rolled her eyes and looked at me as if to say "What can you do in the face of such ignorance?"

Later, in Gaithersburg MD, I met many other interesting people for the first time including the Apollo 14 astronaut Edgar Mitchell, who autographed a copy of his book for me, and the sole surviving witness to the Roswell crash event, Jesse Marcel Jr, with whom I had dinner and struck up a long but distant friendship. He was by that time living in Montana where he practised as an otolaryngologist—an Ear, Nose, and Throat specialist. I was also pleased to see that Richard Dolan was there. And, last—but most importantly—my future wife, Janis.

Contact at Last, and the First Visit

Following the discovery of a scoop-mark scar in late August 2008, I sent photos to Peter Robbins to ask his opinion. This, it turned out, initiated the chain of events which finally led to the contact between Hopkins and myself. Robbins forwarded the photos to Hopkins. Serendipitously, Breakthru Films needed a couple of scoop-mark cases for their film project, documenting the life and work of Budd Hopkins. This led to a telephone call from Budd and subsequently to an invitation to visit in December that year.

After the first call in early September, I heard nothing from Budd for a couple of months and was beginning to think the visit had been delayed or worse, wasn't going to happen at all. Then out of the blue in mid-November another call from Budd, this time asking if I could confirm a visit in three weeks' time. Jan and I returned to NYC during the week of 8th December, scheduled to meet with both Budd and the film crew for several days (at JFK

Budd, Leslie & Steve in Manhattan, April 2009.

Arrivals, we were greeted by the film and sound crew recording the occasion as might be the case for a film star or leading politician, the only time that has happened to me). Budd kindly offered us his apartment for the duration of our stay to spare us the sky-high Manhattan hotel prices, and he stayed around the corner in 15th Street for the duration with his partner, Leslie Kean. Leslie is a successful professional investigative journalist who at that time was researching and writing the manuscript for her forthcoming book, UFOs: *Generals, Pilots and Government Officials Go on the Record*, which caused quite a stir.

Every day of that week was taken up with filmed interviews and memory recovery sessions, a visit to a dermatology clinic uptown for examination and biopsy of the scar, and meeting a seemingly endless stream of people. We were out to dinner with the Breakthru Films team every evening and Budd, despite his compromised energy levels due to his illness, was on top form.

During the less frantic hours of the stay, Budd and I began to form an unlikely friendship. Separately from the abduction issue which had brought us together we had generally similar tastes in art, though his deep knowledge of and passion for the subject far exceeded mine. But I loved and admired the French Impressionists, whose innovations in style were present in the DNA of the NYC Ab-Ex movement. He was the world's best art gallery companion, bringing a passion and love of the subject which brightened the soul. As an amateur art collector I had already acquired an original Hopkins abstract painting from 1976 (a lucky find as not many come up for sale on the international market) plus a limited-run print of one of his circle portraits in blue and yellow, from the 1980s. During our growing friendship over the subsequent years, I was to acquire many more of his wonderful works, including three of his large *guardian* sculptures. The predominantly red-and-yellow *guardian* collage which graces the front cover of his memoir, *Art, Life and UFOs*, now hangs in our living room. He also signed and dedicated all my hardcover copies of his books, all of which I had crammed into my suitcase.

On this trip to Manhattan we also visited the site of the infamous November 1989 Brooklyn Bridge abduction, which Budd had investigated and written about in *Witnessed*, and I met 'Linda Cortile' (not her real name) who was the abductee in that incident. At my request, Budd also called David Jacobs in Philly and arranged for us both to go down there to spend some time with him. The fortuitous consequence of this meeting is described in the next chapter.

Subsequent Visits

Budd hosted me four times as a guest in his Manhattan home over the next couple of years. In October 2010, Budd's partner Leslie was away in Europe for the week, filming interviews in support of her book, and kindly loaned me her apartment. Budd and I subsequently spent a memorable eight days together.

We enjoyed often wide-ranging discussions on art, music (he was a jazz enthusiast with an encyclopaedic knowledge of the subject), history and geopolitics. As already mentioned, Budd had developed progressive/liberal views in his early 20s, in often heated opposition to members of his now-distant WV family, and he was a committed Democrat. He considered the election of Barack Obama in 2008 to be the greatest event in the USA's history, after its founding. He railed against the right-wingers in the US, who obstructed everything that they could of Obama's progressive agenda, in knee-jerk fashion, even as their own states reaped the good fortunes that the federal government offered.

We also discussed history, global geopolitics, many of the abductees featured in his books and, most of all, his acquaintances with many prominent figures in

On the sofa at W16th Street, under one of Budd's Guardian artworks, May 2011.

the art and UFO fields. He had known the late Dr Allen Hynek, Laurance Rockerfeller, the late Carl Sagan of Cornell, and of course John Mack from Harvard. He knew by then seemingly every writer and researcher on the UFO subject both living and dead, including Brits like Timothy Good and Nick Pope whom I had also by then become acquainted with in London. Indeed, Nick was a regular visitor to w16th Street during this time. It was shortly thereafter that he met and married an American anthropologist, and moved from South London to Arizona. Budd was a particularly close friend of Jerry Clark.[17]

Budd allowed me to record many of our conversations but requested they not be made public during his lifetime and in some cases (especially where he was less than complimentary about people, or where the conversation concerned family members) that his comments should never be made public. I have respected his wishes on this and always will.

He showed me around Manhattan and Central Park in brilliant May sunshine. We lunched more than once in the quirky indoor Chelsea Market, visited art galleries, and had engaging conversations while walking the wonderful High

Line, the first section of which had just been opened as an urban garden with panoramic views over the city. Budd had lived in New York City for more than fifty years; there was little he didn't know about this wonderful place, which he loved dearly. He once told me that he had observed, among his New York City friends, that "Musicians are nomads, but artists are nesters."

Undertaking abduction research is a guaranteed cash drain. As mentioned above, Budd never once charged anything to any abductee, while generously giving his time and resources to anyone in need of support. As the consequence of a successful art career—and moving to New York City in the 1950s before property prices went crazy—Budd was definitely asset-rich: the 16th Street house was worth multiple millions of dollars by this time. But he wanted to leave it all intact to his daughter and granddaughter, thus he was not as cash-rich as he would have liked. I had arranged for a wealthy business contact then living in Hong Kong to visit the studio to buy some artworks, as he—I'll call him Jack—was a successful financier and collector of art, and had a permanent home in upstate New York where he'd planned to keep his collection. Jack was also interested in the UFO and abduction subject and had quizzed me about it many times, genuinely intrigued by my experiences.

Budd was away from home during the week Jack planned to visit, so Peter Robbins graciously offered to show him around the works available for sale. Jack made his selection, then dashed off back to Hong Kong. I later rented a car and we loaded the artworks, then drove up to Jack's private home in a small community north of the city to deliver them.

It was during this 5-hour road trip that Budd revealed to me that in recent years art critics, gallery curators, and fellow artists had told him that his fascination with the abduction subject and "all that weird UFO stuff" had damaged his reputation in the art world and distracted people from his achievements. He had been advised to drop the subject and re-immerse himself in the art scene in order to restore his reputation. *Art, Life and UFOs* touches on this theme: "*There was confusion in the art world as to what it was that I did in life: was Hopkins a painter, an art writer, a UFO investigator, a journalist, a teacher, or what?*" [18]

Budd, however, always remained committed, as he considered the abduction phenomenon as being too important a subject. Having opened the door to a continuing avalanche of letters and messages from abductees, he felt a strong responsibility to them. He was still receiving ten or so letters each day in 2011, the year he passed on and ten years since he published his last book on the abduction subject.

Reflections On the Man

In the final year of his life, as he was assaulted by lymphoma and secondary cancers, Budd kept a journal. He also continued to be a voracious reader. One day, he wrote:

> I am reading Graham Greene's memoir, *A Sort of Life*, which has a little too much about his earliest childhood—games played, children's books read, etc—but he is a marvellous writer, clear and down to earth. He mentions in passing that he found Haggard's famous *She: A History of Adventure* to be a "sloppy, metaphysical love story" and then adds, in parenthesis, by way of explanation, "I have always preferred Freud to Jung." I sense a kindred soul. If I extrapolate from his remarks about H. Rider Haggard to those of us doing research on the UFO issue (and life in general), Dave Jacobs and I vastly prefer Freud to Jung while John Mack was a Jungian. This distinction, I believe, explains a lot. It defines the difference between precision and cloudiness, between soil and smoke, between earthly humanism and unbounded mysticism.[19]

Having read John Mack's 'interesting' take on the subject in *Abduction: Human Encounters with Aliens* and *Passport to the Cosmos*, I found this passage to be most astute. We'll return briefly to Dr Mack and his work with abductees in Chapter Eight.

Budd had a joyous and mischievous sense of humour and could be hilarious when telling a risqué story or joke, often spiced with lively theatrics. I feel certain that he would have been a fine stage actor. One of my warmest memories is of him dancing around his apartment, age seventy-eight, with his infant granddaughter GiGi, when he demonstrated for a brief time that he could be almost as energetic as she.

One small task I was able to perform for him was to attempt to sort out and organise a pile of several hundred unopened letters. Administrative organisation does not come naturally to me but, having run a successful business for eighteen years, I had developed some experience in performing these essential tasks with reasonable efficiency.

The predominant volume of correspondence concerned the Intruders Foundation and included dozens of cheques from members and supporters. There were also household bills for the Manhattan property; bills and correspondence for the Wellfleet property, which Budd and April time-shared

With Budd & Randall Nickerson in a Lower West Side diner, 10 May 2011. This was to be our last meeting, as Budd passed away on 21 August.

through the summer months; general correspondence for IF which consisted mainly of people recounting abduction experiences; correspondence on art gallery matters; and finally some personal correspondence. I did the best I could sorting it all out into piles under different categories, especially returning the cheques to IF where return addresses were supplied, or simply destroying them, as Budd wouldn't take the money. At this stage in his life, he had neither energy nor appetite for any of this and wished to spend his remaining time engaged with things which interested him, like reading books and conversation with close friends.

As I left Manhattan later that week, I knew that the next day more mail would arrive and begin to pile up. However, London was calling and I needed to get back.

One small incident is worthy of note. Prior to having left London for New York, I had directed a company in Utah which produced herbal medicines to mail a parcel to Budd's home. I had planned to take it back in my luggage, to avoid both delay and import taxes. I called Budd from England before departing for NYC to check whether it had arrived, and he told me the parcel had arrived "and it's sitting on one of the kitchen chairs, like a lonely houseguest waiting for dinner." Upon arrival, I opened the package only to discover that it was not my herbal medicines but a heavy artefact: a "Lifetime Achievement Award to Budd Hopkins," about the size and design of a Film Academy Oscar, but from a UFO organisation. This was not the first time he had been the recipient of such formal public recognition.

Besides the Breakthru Films team, another regular visitor to Budd's home during this time (2008–2011) was the radio and podcast host David Biedny. He was also teaching computer science at MIT and brought around one of the pre-market launch Apple iPad demo models for us to play around with. The touch screen and fingertip control was revolutionary. Familiarity with the now ubiquitous tablets breeds the kind of complacency common to the human brain when tech is accepted and embraced so quickly and widely; it seems like they have always been there, but they haven't.

Another regular visitor was Randall Nickerson, an abductee who, with Dr John Mack, had been interviewed in the late 1990s by Oprah Winfrey. Randall had recently returned from a six-month trip to southern Africa with his camera and sound gear, where he'd been following up on the now-adult witnesses to the 1994 Ariel School UFO landing incident at Ruwa, Zimbabwe. The trip had been an eventful one. While seeking out the original witnesses, Randall had been thrown in a squalid and overcrowded jail for a time by the Zimbabwe police. He also stayed at a place (in Botswana if memory serves me well) where a large number of venomous snakes were kept on the premises and he was advised to place his bed in the centre of the room "because the snakes stick close to the walls as they move around." It didn't sound like my ideal kind of holiday. Randall was in the process of editing the 300 hours of interviews for the documentary, *Ariel Phenomenon*, a task which consumed almost all his available time.

Randall was also a qualified pilot and a skilled musician: after John Mack's death in 2004, Randall had played piano at the memorial service. In October 2011, he would do the same for Budd.

In his final year, Budd's health declined to the point where he became less able to travel long distances. Yet he retained his sharp mind and his sense of humour despite the pain he was in for much of the time, and his more general decline in health and stamina. He wrote in his journal on 7th December 2010:

A tiny consolation from E.M. Forster's *Howard's End*:

"Even if there is nothing beyond death, we shall differ in our nothingness."[20]

Budd Hopkins died on 21st August 2011, two months after his 80th birthday, from a combination of lymphoma, liver cancer, and pneumonia. He was a humanist to the last and skeptical about any kind of after-death survival, but the reader is urged to read Leslie Kean's truly excellent second book, *Surviving*

Death, to assess whether Budd Hopkins may have indeed been confirmed in his belief that "... there is nothing beyond death," or whether, "on the other side," he received a welcome surprise.

In October of that year Grace and Leslie organised a memorial event for Budd in Manhattan which was attended by more than 400 guests. In addition to family, friends, and the extensive art community in New York City, the occasion brought together almost the entire research community from the UFO field, including several who travelled from Europe and Latin America to honour the great pioneer of abduction research.

Rest in peace, my good friend.

We'll next take a look at my work over several years with Dr. David Jacobs, and how he opened up my understanding of the phenomenon, making it somewhat less troubling and mysterious. During this next chapter, we'll look at memory recovery hypnosis: what it is and is not, and how it works. We'll also examine exactly what happens during the 'mindscan' procedures to which all abductees, so far as is known, are routinely subjected during virtually every abduction event.

Chapter Seven

The Assiduous Professor

*My invaluable work with Dr David Jacobs
and our growing personal friendship*

I first became aware of David Jacobs and his abduction research in the 1990s, although had little idea at the time that this phenomenon had featured so persistently in my own life. The precise time and circumstance now evade memory, but I clearly remember Dr John Mack from Harvard University giving an interview to BBC Radio 4 about his work with abductees in his clinical practice. I'm unsure whether this interview coincided with the publication of one of John's books or something related to his problems at Harvard, but Dr David Jacobs was mentioned by Dr Mack as being a fellow academic also involved in the ground-breaking study of this phenomenon.

This narrative has already mentioned that Dr Jacobs and I first met face-to-face in December 2008 when we visited him at his home in Philadelphia. However, we did have a brief interaction via email a year prior to that. We'll come back to that later but first of all, who was this guy and what is his back-story?

Brief Bio

David Michael Jacobs was born in Los Angeles CA on 10th August 1942 and spent part of his childhood in the Sunair Home for Asthmatic Children.

Highly intelligent and with great academic aptitude, the young David Jacobs exhibited an early interest in the UFO phenomenon and joined APRO, Jim and Coral Lorenzen's Aerial Phenomena Research Organization, as an amateur field investigator of sightings in the 1960s. When asked how and why he became interested in the subject (Had he ever seen a UFO? No, never …) he simply responded, "I just became interested in it, that's all." It seemed to Jacobs that it would be of enormous importance if it proved to be 'contact' with an

First meeting with Dave Jacobs in December 2008.

advanced and obviously technologically superior civilisation from another planet or star system, and at the time this was the eventual outcome most people expected and hoped for. Back in the 1960s the UFO issue was less marginalised in academia with committed PhDs and STEM scientists of every discipline serving on the advisory boards of such organizations as APRO, NICAP and MUFON (more than 200 accredited academic scientists worked officially with APRO), and was not the marginalised, kooky fringe subject it later became.

Jacobs completed his BA in History from UCLA (University of California, Los Angeles) and then moved on to the University of Wisconsin, Madison, where by 1973 he had completed both his MA and PhD. His doctoral dissertation was titled, *The Controversy Over Unidentified Flying Objects in America*, a comprehensive analysis of the way the UFO phenomenon had been handled by the military, Air Force and Navy, successive government administrations, the intelligence agencies and the mass media. Although there has always been a small cohort of engineers and scientists who took the matter seriously—those who joined the likes of NICAP and APRO, for example, or had managed to talk the leadership at one or another major aircraft manufacturer into funding discrete studies of the subject—academia has never been overly welcoming to 'flying saucers'. Jacobs' PhD was only the second related to UFOs that had ever been awarded in the United States.[1]

The dissertation was subsequently published, in a slightly revised version, by Indiana University Press in 1975. It sold out in hardcover, resulting in a second print run—an unusual level of popular success for any academic work. The work by the 33-year-old Jacobs was described by Arthur C. Clark as "one of the few volumes on the subject which is worth reading." This was an accolade indeed, as Clark did not believe there were any *unidentified* flying objects and all sightings had a prosaic explanation. He'd once written that people reporting UFO sightings "tell us absolutely nothing about intelligence elsewhere in the universe, but they do prove how rare it is on Earth."

After a spell as an assistant professor at the University of Nebraska, the young Dr Jacobs secured tenure at the Department of History at Temple University in Philadelphia. He was to spend the remainder of his career at Temple before retiring from academic life in 2011, aged sixty-nine.

Dr Jacobs offered the only curriculum university course on UFOs in the United States, titled *UFOs and American Society*, which was offered for twenty-five straight years. Thousands of university students learned formally about the UFO phenomenon within the discipline of academia through Dr Jacobs' course. David told me that many students over the years asked how they could become involved in abduction research. His advice was always: "Don't even consider it or think about it until you get tenure. In the meantime, don't mention the subject to any prospective employer."

With a keen interest in the UFO phenomenon, Jacobs was obviously aware of the first reported abduction cases through the 1950s, '60s and '70s but like most researchers, considered this aspect of the subject too outlandish and improbable to take the reports seriously.

In an appendix to his 2015 published work, *Walking Among Us*, titled, "Evolution of an Abduction Researcher," Jacobs explains his attitude to the abduction phenomenon through the early years when the first reports gained wide publicity:

> In the beginning, I placed little stock in the abduction phenomenon. I found it interesting, but probably psychological in origin. The use of hypnosis, a problematic technique, did not lend it legitimacy. Most abduction evidence is the result of human memory, with all its problems, retrieved through hypnosis, with all its problems, administered by amateurs like me. It is difficult to think of a weaker form of evidence, especially for such a potentially important subject.[2]

His curiosity was deepened, however, on meeting Budd Hopkins in 1982. Hopkins, Jacobs discovered, was both highly intelligent and rigorous in his methodology when investigating and analysing abduction reports. For the first time in more than twenty years investigating and reporting on sightings, through looking at the work of Hopkins, Jacobs began to be genuinely intrigued by the abduction accounts, and he realised that they ought to have been given more serious attention from the beginning. He remembers calling Allen Hynek and commenting on how impressed he had been when confronted by Hopkins' methodology and accumulated evidence. Hynek's advice was that the subject was considered just too weird: "Don't go anywhere near UFO abductions, David." For Hynek was well aware that, for all too many, Judge Frankfurter's response applied: "I am not saying that the witnesses are not telling the truth, but *I simply cannot believe it.*"

To his credit, Jacobs was not deterred. He spent time at Budd's summer house in Wellfleet on Cape Cod, where he met with several abductees, and watched Hopkins' careful methodology in action. With the Philadelphia area correspondence from among the piles of letters in Budd's office, he returned to Philadelphia determined to start working with local abductees.

He started the work in earnest in 1986, continuing in the methodology of Dr Benjamin Simon, who had treated Barney and Betty Hill in 1961. Under Hopkins' direction, who had himself been mentored by two New York City psychologists, Jacobs worked at these techniques with patience and persistence. The results of his first five years' work, *Secret Life*, was published in 1992. In this book, for the first time, a researcher was able to describe, step by step, what happened during a typical abduction event and, moreover, describe:

1. Primary procedures which are almost always carried out on almost all abductees

2. Secondary procedures which are carried out only sometimes on most abductees

3. Ancillary procedures which are rarely reported but to which some abductees are occasionally subjected

Diligent and attentive work led to several other realisations about the abduction phenomenon. Of primary importance was the ability of the abductors to pacify abductees from a distance—perhaps as far as several hundred metres—and to render unconscious any potential witnesses in proximity to the scene of an abduction.

'Mindscan' and Staring Procedures

Jacobs also realized that the abductors possessed both a deep knowledge of human brain function and the skill to manipulate neurological processes to such a degree that they are able to bury the memories of each event very effectively in the abductee's long-term memory. They do this via a 'staring procedure' during which a 'mindscan' (a word which Jacobs no longer favours) is deployed, usually by one of the taller greys.

Abductees reported that, while lying down and immobilised, the alien would stare into their eyes from only a couple of inches away. Jacobs questioned them as to what had been going through their minds as this was occurring, eventually working out that the alien was skilfully manipulating the mind and memory of the abductee. It has since been assumed that this memory manipulation is essential to maintaining the secrecy of the abduction program. Abductees are left with a gap in their memories, termed 'missing time'.

In order for the investigator/therapist to overcome this block it is necessary to develop questioning skills which avoid imparting any suggestion to the subject. Conscious memories, Jacobs and others have learned, are notoriously unreliable, with bits and pieces of memories often assembled by abductees in any random order, rarely corresponding to the reality of what occurred and with confabulation, assumption, and fantasy liberally stirred into the mix.

This situation is made even worse by the abductors' abilities to place screen memories and images in abductees' minds, with the purpose of intentionally deceiving them about what happened during their abductions.

Abductees reported that this staring 'examination', usually carried out with the heads of the abductee and the alien almost touching, takes several minutes and occurs every time they are abducted, without exception. Abductees describe awareness of their brain patterns being accessed and manipulated during this procedure, experiencing actual physical sensations in the internal organs of the body with emotions of euphoria, anger, fear or sexual desire frequently aroused.

Thanks to his diligent, careful work with more than 150 abductees over thirty years, studying the anatomical neurology in some medical detail, David

Jacobs gradually worked out what might be going on here. No one else at that time got even close to understanding or being able to explain this process. From his book, *The Threat*:

> The aliens' ability to stare into an abductee's eyes and effect a wide variety of changes in brain function is extraordinary. At first it seems almost supernatural or mystical, but the mystical and supernatural are not part of the abduction phenomenon. The aliens use their advanced knowledge of human physiology to control them and ensure that the alien agenda is complied with and secrecy maintained.
>
> The aliens' ability to control humans comes through the manipulation of the human brain. For example, when the alien moves close to the abductee's eyes to begin the staring procedure, almost immediately the abductee feels emotional and physical effects. One way to explain this is that the alien uses the optic nerve to gain entrance to the brain's neural pathways. By exciting impulses in the optic nerve, the alien is able to travel along the optic neural pathway, through the optic chiasma, into the lateral geniculate body, and then into the primary visual cortex in the back of the brain. From there he can travel into the secondary visual cortex in the occipital lobes and continue into sites in the parietal and temporal lobes and the hypothalamus. Through that route, the alien can stimulate neural pathways, travel to many neural sites and cause the firing of neurons at whatever sites he wants.[3]

Jacobs then describes how the abductors can manipulate memory and cause abductees to clearly remember events that did not happen:

> Brain stimulation allows the alien to produce a range of effects. If the alien can connect to the neural pathways, he can reconstitute an abductee's memories. He can inject new images directly into the visual cortex, bypassing normal retinal observations and cause people to 'see' things that become part of their abduction 'memories'. He can activate sites within the limbic system and cause strong emotions, such as fear, anger and affection. He can create sexual arousal that builds relentlessly to a peak. And he can institute a form of amnesia that helps to preserve secrecy.[4]

Then the process is outlined whereby the taller alien may achieve this level of entry into the brain patterns of the abductee. It was explained earlier in this narrative that abductees normally retain control over the movement of their eyes, but during the 'mindscan' procedure this, too, is tightly controlled:

> By using the optic nerve the alien can, in effect, travel down the brain stem into the autonomic nervous system in the spine, and then branch into the parasympathetic nervous system, giving him contact with virtually any organ. Abductees often talk about feeling physical sensations in their genitals, bladder or other areas when an alien performs mindscan procedures. The physical responses responsible for erection and ejaculation in men, and for tumescence, lubrication and expansion in women can be artificially generated in this manner.
> How the aliens engage the optic nerve is, of course, unclear, but there are some clues. When mindscan or any staring procedure begins, the abductee cannot avert his eyes; they must remain fixed and open. The abductee is, in effect, forced to peer into the alien's eyes. Most abductees report that his eyes are dark brown or black, and opaque. Others describe what might be a liquid linside the alien's eyes. Others frequently see a wiggling structure in the back of the eyes that generates a "light". It is possible that the light-emanating mechanism engages the optic nerve to begin the alien's journey through the neural pathways.[5]

Jacobs quotes the recorded testimony of several abductees describing these procedures (the names and identities are disguised):

> Some abductees can feel the engagement when it happens. Alison Reed often felt the alien's physical attachment to her brain during mindscan:
> What's he doing when he's inside there?
> I feel a little tired. There's that thing again. I can't see it but I can feel it, its ... and it goes all around, it's like a blue light. It's between my skull and my brain, of course I can't see it, I just feel it. I don't feel much of anything right now. I feel good, I feel relaxed ...
> The blue light, is that from his own eyes, do you guess, or from an instrument?
> No, I don't like to call it a light because it's not like a light you see, it's more like an energy. I can't see it, usually in these places you see certain things but you feel more than you see. Your major senses are no longer sight and smell and touch, it's your sixth sense when you're here. It's from him, it's not an instrument, it's an energy. Somehow he can make this energy go in my head.

Similarly Courtney Walsh, a young woman pursuing a career in the biological sciences, felt her neural pathways being stimulated.

No, it feels like, it's hard to describe, like something is worming around in there. You can feel the different nerve pathways ... it actually feels nice though. I can feel actual—it feels like something is—little currents of energy running around in my head.

Jack Thernstrom, a graduate student in the physical sciences, had a similar reaction and sensed that the alien was physically going through his mind.

Now he's looking in my face again, and this time it's that feeling of a knife prying into my mind.

This is a feeling of ... a physiological situation that's going on there?

It's like pure mental pain.

What do you think he's doing now?

I have this impression of, as if he's probing his way through a lot of—it's almost a physical sensation, as if thin strings or cables are all closely intertwined, almost hairlike, but under tension. It seems I've seen something like this ... he's kind of groping in there, and finding paths in there to get to a certain point. It's this feeling of a knife probing through, and forcing its way between things ... it's somewhere between active and passive ... it's not like opening it up and looking at it, it's as if one had a mass of wires and one were pulling and separating them to see what's connected to what.

Some abductees visualise random thoughts and images as the alien traverses the neural pathways, as if the travel enervates the pathways as a by-product of the procedure ...[6]

Most importantly:

Once joined with the abductee's neural pathways, the alien essentially has free rein to do what he wants. The abductee is no longer in control of his own thoughts. The alien can exercise absolute power over the minds and bodies of abductees. They can make the abductees think, feel, visualize, or do anything the aliens want.[7]

And the techniques involved would seem to indicate this access is sometimes not without difficulty for the alien:

The aliens' abilities to attach to the abductees' neural pathways is not automatic. They turn and twist their heads to get the best vantage point to hook into the optic nerve. They hold the abductee's head so that she will not make any movements that might disrupt engagement. Karen Morgan had an unusual mindscan in which the first alien could not make an adequate attachment. After the first alien tried without success for several minutes, another alien took over and she could quickly feel the effects of the familiar mindscan procedure.

But another abductee successfully resisted engagement. During a recent abduction, RK found she had more muscle control than usual and she used it to prevent neural connection. She shifted her eyes back and forth rapidly while reciting an Arabic religious phrase [this abductee is a Moslem citizen of the USA]. The first alien tried to lock into her eyes but could not. He diverted her attention by causing a pain in her head and threatened not to take her home, but she refused to give in. Another alien took over and increased the threats. Still she refused to stop, although she was getting dizzy moving her eyes back and forth. A third alien tried, then a fourth. They could not stop her from shifting her eyes. Eventually they gave up and said they would continue the procedure at the next abduction.[8]

Jacobs' careful, diligent work with abductees might very well explain the source of, and reason for, the *missing time* phenomenon. At the end of the abduction, abductees do not remember anything about what just happened to them and have a seamless gap in their memory, often with several hours of memory just missing:

> Although the exact neurology is not known, it is most likely that the aliens store the abduction events directly in the abductee's long-term memory system, bypassing short-term memory and preventing the triggering mechanism that allows for its reconstitution. Hypnosis restores the trigger that allows the memories to come forth. RK was told that the reason the aliens do not erase the memories altogether is that there are aspects of them that must be retained by abductees for future reference. Thus, the memories are intact but inaccessible to normal recall.[9]

Why do the abductors perform these procedures?

One obvious assumption is that if the aliens can ensure the abduction event is stored only in the abductee's deeply buried, long-term memory, then secrecy about the program can be maintained as the abductee will only be aware of the seamless missing time—and not even that if, for example, the abduction takes place when the person is sleeping.

However, there may be other reasons too. Jacobs again:

> Abductees have said that in some way they know the mental procedures are related to the hybrids. The abductees suggest that aliens record information from them and then transfer it into hybrids' minds so they can learn how humans live and feel. There are also procedures in which hybrids directly transfer information from humans directly into their minds. An alien attached Alison Reed to an adult female hybrid and as the two sat facing each other, Alison could feel her thoughts and memories flowing out of her and into the hybrid. The hybrid absorbed Alison's thoughts and experiences and apparently derived some benefit from this procedure.
>
> The mental procedures must be viewed in relation to the aliens' reproductive agenda. Without the ability to manipulate the human brain, the aliens would be unable to control the abductees physically or mentally and the breeding program would not be feasible in its present form. Abductees often feel more violated by the mental procedures than by the reproductive ones. They know that their private thoughts are not their own and that they can be "tapped into" and manipulated. Although I often try to reassure them that in spite of what happens their thoughts are free, they know that this may not be entirely true.[10]

All this time, Jacobs was a popular and successful professor at Temple University,[11] highly respected by his students. But when his research into alien abductions became more widely known his academic career began to be adversely affected. Despite the popularity of his books on the subject, which the university acknowledged by awarding academic credits, and positive media coverage, the writing was on the wall. He had tenure so his position was secure. He was popular with the student body, respected as a tutor and capable historian by professional colleagues, and his curriculum covering twentieth century American politics and society was popular and successful. But the university authorities did not want to be associated with kooky alien abduction stuff, and he was informed that it was not to remain among his professional activities. Like John Mack, Jacobs paid a price for his commitment to the subject.

Around 2009 I asked him why he had not taken a year out to publish a research work on some other aspect of modern American history, which might have diluted or moderated the hostility he endured due to his focus on the abduction work. He responded that he considered understanding the abduction issue so vital for the future that he would not be distracted from it despite the obvious damage it was causing to his career, and it would remain his primary focus, though outside his faculty responsibilities at the university.

First Brief Exchange of Ideas

In the latter part of 2007 I had read all four of Dr Jacobs' then-published books. On reading *Secret Life*, I was genuinely astounded that this man seemed to be inside my head, as though he'd been observing my personal life for many years and knew everything about me (he hadn't, of course, but became aware of abductees' lives and what happens to them as a result of studying the subject). I completed one of his questionnaires from the now-redundant ICAR website, and later followed up with an email, as I had a specific question about the intergenerational aspect of this phenomenon. What follows is the background to that enquiry.

I was the firstborn child of my mother. Apart from the fact that she was an abductee should have been obvious to anyone who knew her and was familiar with the phenomenon, she attempted to explore the issue of what conceivably might be the causative agency of all the weirdness which afflicted us with me several times during the final years of her life, but to little result as I did not then understand the subject at all. She died suddenly in September 2000.

She was a twin, born thirty minutes prior to her twin brother. Thus, she was also the firstborn. Her mother was also firstborn, an only child born in 1908. When I asked about the night-time abductions in 1967 during communication with one of the taller greys, it responded cryptically that, "it's because of your mother, and your grandmother, and great-grandmother."

My great-grandmother was born in 1873, so may have been one of the original abductees selected for the program. She was the only one of my eight great-grandparents I ever knew: she died in January 1963, when I was six. She was the eldest of three children, though I am not sure how relevant this might be if she was one of the 'originals'. We'll explore some origination theories about the program in the final chapter.

My question to David concerned this aspect of the phenomenon, as I was not convinced some of my other relatives in this line were abductees. We had a dialogue about that. He was convinced that all the children of abductees will themselves become abductees. Although I accept that he has sound reasons for this conclusion, I remain unconvinced on this point. He is probably right though, as he has proved to be right on so many aspects of the phenomenon: I just don't see it in my own family.

December 2008, and an Offer I Couldn't Refuse

We finally met in person in December 2008 when I visited David and his wife Irene at their large house in a Philadelphia suburb, where they had lived for decades. Their two sons having recently 'fled the nest', they now shared the house only with their two cats.

David's studio and library occupied almost the whole of the top floor and contained thousands of UFO books and journals collected over decades of study. He and Irene were revealed to be highly discerning collectors of twentieth century art and music. When added to David's academic expertise in American history, the house was a treasure trove. They had an elevator which rose from the grand hallway, an unusual and eccentric feature installed by a prior owner of the house. David and Irene proved to be delightful hosts, both of them intelligent and good-humoured.

This was the first time I met David Jacobs. A gold-plated endorsement from Budd Hopkins was sufficient for David to offer to work with me on my experiences. Meeting in person clinched it. I would be fortunate to deal with him much more extensively during the following years as he generously consented to work with me to uncover what had been happening with "our little friends," as he drily described them.

Getting to Know the Man Himself

Over the thirty-two years during which he worked with abductees David worked with just 150 individuals, though he was extremely thorough. (For comparison, Budd Hopkins worked with more than 1,000 people between 1977 and 2011.) Devoting as much time as the abductee needed or requested, an average session looking into a single abduction event with Dave Jacobs might last up to five hours. I myself was one of the extremely rare subjects who worked with him remotely, over Skype, as regular attendance at his office in Philly was not possible. Once he agreed to work with an abductee he gave them

With Dave Jacobs in the rain, Philadelphia April 2009.

as much of his time as they required. Some abductees worked with him continuously for more than twenty years.

David's work with me was invaluable, and I shall always be grateful for all the time and effort he devoted to me. We became friends, and regularly phoned or skyped to chat about all manner of things. In the summer of 2013, for their wedding anniversary, David and his wife Irene took a long vacation to Europe, mainly in London and Paris. They took the opportunity to visit us at our Hertfordshire home, taking the train from central London to our country village 25 miles from the city and visiting our local gastro-pub with us. One time in Philly, David gave us a brief tour around the university and Independence Hall, where the US Constitution was ratified. He conducted us around the National Constitution Center and the Liberty Bell, which I had had never realised was cracked. On this trip to Pennsylvania we also visited Gettysburg and Lancaster County, where many Amish live their traditional lives uncluttered by modernity. On his visit to Cambridge UK in October

2016, we drove up there to spend the day with him and his friends who lived in the city and with whom he was staying. We met for a second time in Cambridge for an extended restaurant lunch in 2018. At this time, the truly remarkable Professor Stephen Hawking was still lecturing in astrophysics at the university and could be seen trundling around the campus in his motorised wheelchair.

Most people who only know *of* David Jacobs—but never actually *knew* him personally—do not and cannot know what a great guy he is. He's warm, good-humoured, and delightful company with a keen interest in the arts and sciences. An excellent and always stimulating conversationalist, he has an encyclopaedic knowledge of political history, art, and cinema. He is always even-tempered, has a radiant, optimistic disposition and a wry, self-deprecating sense of humour. He cares deeply about the abductees who work with him and is always available for them if they call.

The Careful Process Leading Up To Regular Hypnosis Sessions

So, how do you recover the buried memories? The answer is that 'hypnosis' can do it. No one knows how, or at least I don't, but it does work. But not necessarily as you might think.

A lot of rubbish has been talked about hypnosis. The way Dave Jacobs always used the technique was to encourage the abductee to relax and to answer some simple, logical questions about what happened, starting with the time before the event and moving logically from one thing to the next. There is nothing magical or occult about the process, but it does take time to learn and practice and care needs to be exercised in certain areas. The questioning needs to be careful, as it's vitally important not to offer any suggestions to the subject, as it's vital to allow sufficient time for the memories to emerge. You are not under any kind of 'control', you are just talking to someone normally and can get up any time and go to the bathroom or answer the door, or even make some coffee, and then reconnect and carry on with the session.

Claiming that you should be a qualified psychiatrist to do this is exactly the same as saying that only a degree-qualified automotive engineer should ever be allowed to drive a car. Driving is just a set of learned skills, ditto hypnosis for memory recall. It takes practice and needs to be learned, like driving a car, but nearly anyone can learn to do it. Many abductees learn to do it for themselves, with no practitioner necessary. Budd Hopkins was ultra-cautious during the first seven years he was investigating abduction accounts through the late 1970s and early 1980s. He'd arranged for professional Manhattan psychologists to conduct the hypnosis sessions and subsequent interviews, with Hopkins not even present for any of these interactions. It was only when these same professional psychologists persuaded him that relaxation hypnosis was a simple set of skills which anyone could learn that he allowed them to mentor him in learning how to do it.

Prior to meeting with Dave Jacobs, in 2007 I had visited two different hypnotherapists in the UK to try to access the memories. Both were qualified and working in the National Health Service, though one also had a stage-hypnosis business which was by his own admission far more lucrative than working in the Health Service to assist people to stop smoking or lose weight. The other was recommended to me by Nick Pope, with whom he had been a lifelong friend since childhood. They were well-intentioned guys and worked with abductees for free, but neither understood much about the abduction program so didn't know the right questions to ask, so we achieved little.

However, it was after working with one of these hypnotists that I first began to experience memory flashbacks during everyday life. The most striking example happened when walking along the quayside at Halong Bay, in northern Vietnam, watching the Sun go down in December of 2007. A very strong memory emerged, right out of the blue, of one of my up-to-that-time-completely-forgotten night-time childhood incidents. It was shocking and completely unexpected, as I was not even thinking about the abduction subject at all at the time of this sudden recall.

The emergent memory was completely real and *suddenly just there*, intact and detailed, after forty years. This was a full six weeks following my last session with the hypnotist, so this process of emergence from the long-term memory can take time. When this happens, you realise that you have always known this and remember it well, and you can't understand why you hadn't remembered it yesterday, or for forty years previously. It's because the abductors are skilled manipulators and very effectively bury the memories *unless and until you deploy a technique to access them*.

Working Methods and Safeguards

Prior to consenting to work with a suspected abductee, back in the day David Jacobs used to ask the applicant to fill out a long questionnaire. His International

Center for Abduction Research website is still online, though no longer maintained. However, one can at least visit to see how David managed and pre-screened abductees prior to consenting to work with them.

Having filled out the questionnaire, it was then up to the applicant to follow up and initiate contact. David at this stage would engage with the applicant and warn him/her of the risks of continuing. These are, basically, that when you 'open the door' on these experiences you may find you don't much like what's on the other side. There is no turning back once you remember what the abductors have buried. It is important that the individual who suspects s/he may have had abduction experiences is ready to initiate this process, and not unduly preoccupied with other life matters which demand attention or be temporarily consumed by other responsibilities. Suddenly remembering these experiences can be disruptive and disorientating, and you need a clear mind and space and time enough for the resulting adjustments to your reality to take place without being distracted by other life concerns.

Jacobs thus would try to put people off, and it was only with those who were proactive and persistent that he continued the process. He encouraged a waiting period of two or three weeks during which the abductee should think things over, after which they should call him *yet again* if still wishing to proceed. The only contractual arrangement that I recall was granting the right for David to record and report the revealed experiences, ensuring the subject's anonymity. David has never compromised anyone's identity in thirty-two years of these arrangements.

Furthermore, David agreed to never place any part of the recordings or transcriptions in the public domain without the expressed consent of the abductee (never compromised by Jacobs even once during the thirty-two years), and was scrupulous about providing an unedited recording of every session to everyone who worked with him. The self-suspected abductee requesting that Jacobs work with him/her similarly contractually agreed to never place any part of the recorded conversations in the public domain without Jacobs' explicit written consent. This contractual agreement was violated by just a single individual in those thirty-two years.

Many of my recalled experiences were reported accurately in the 2015 book, *Walking Among Us*, obviously under a pseudonym, so I will not repeat them here. David had more than 1,500 people each year who contacted him requesting him to work with them, and always had a substantial waiting list, but never charged anything to any abductee, ever.

My Experience of Memory Recovery and How It Works

Here are some observations I shall offer from working with David Jacobs on the memory recall between 2009 and 2016. It is important to understand that the individual attempting to access his/her own long-term memories is the fundamental determinant of any resulting memory recovery. With the best will in the world, s/he sometimes doesn't remember much during the first few sessions. However, practice makes perfect. You can genuinely improve recall by refreshing memories of events not visited for years, being careful about the details. Be patient, and with practice, results will come.

The hypnotist's role is confined to offering guidance and focus. A good practitioner will often not say very much, other than 'What happened next?' or 'Uh-uh' every few minutes, and otherwise remain silent. The first session I remembered almost nothing and thought the exercise might not work with me. But I was determined to follow through, and therefore requested a follow-up session.

By the third session, I began to recall a continuous and joined-up narrative and even some procedures which were a shock to me, as I had never encountered them in the literature. During the hour-long debrief, Dave told me that he had heard what I had reported a number of times, including the subject being sat in a contraption resembling a dentist's chair. He still did not understand what some of these procedures were for, however.

By the fifth or sixth session I was able to remember most of what happened, with Dave just listening, and rarely asking me to clarify anything. His involvement in the conversation was confined to the occasional "uh-huh" to reassure me he was still paying attention, and, "OK, what happened next?" as an occasional prompt.

Within a couple of months, I was beginning to experience flashbacks of vivid memories, spontaneously recalled while engaged in everyday activities, or simply relaxing after a day's work. Some of these unexpected spurts of memory were shocking or frightening. On one occasion, I was thinking idly about a business issue and suddenly sat bolt upright with the clear, technicolour memory of a female hybrid/hubrid staring into my eyes. She had black, straight hair with a fringe, green eyes, and was wearing a green dress, smart black shoes, and cheap-looking jewellery. Her skin was pale white and she wore make-up. Unlikely as all that may seem, we were definitely, unequivocally aboard a UFO, and she was engaged with invading my mind and reading my thoughts. Very, very shocking and scary; I was shaking with the sudden trauma

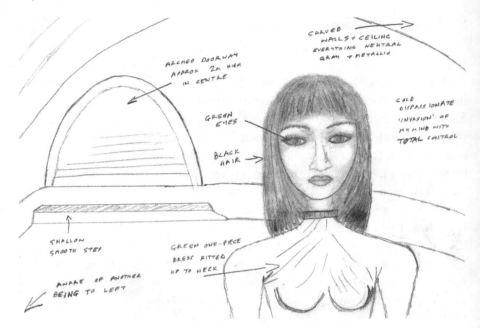

Drawing of hubrid onboard, vividly recalled in 2016 several weeks following a hypnosis session to stimulate memory recovery.

of the memory. The image was so powerful that I proceeded to draw it afterwards. I wish I were a better visual artist and could have done a better job of reproducing what I saw more precisely and in greater detail. Nonetheless, Dave subsequently used the drawing in several of his conference presentations.

When I related to Dave that I was absolutely certain this encounter took place aboard a UFO, and that the hybrid being had been wearing normal, human clothing and jewellery, I rather expected to be challenged. I had neither heard nor read of any such reports before. Commonly, these hybrid beings are reported to be wearing plain, loose-fitting and identical beige- or neutral-coloured casual garments. A thoughtful pause followed before he told me that he had, in fact, received a number of such reports recently, and that what I'd reported was possibly further confirmation of this new phase of their program. In *Walking Among Us*, David related reports of a 'cocktail party' on board a UFO in which abductees engage with 'hubrids' wearing western business attire, learning to act 'normally' when among humans on Earth. It sounds ridiculous, but this is precisely the kind of detailed encounter which many abductees vividly recall.

What is happening with these flashbacks is that the pathways to the long-term memory are being opened by the hypnosis. But the memories emerge at a rate that is not always possible to control. They may not emerge during the session but come later—sometimes several weeks later. Budd Hopkins explained to me once that it's as though your foot is placed on a garden hose to stop the water flow, and when you release the blockage the water spurts out all over the place with the hose jumping around, barely controllable. Only after a time can one gain some control over the emergence of the memories. Dave Jacobs was of course acutely aware of this process and its unpredictable timing, which prompted his cautions during the screening process.

By the time of the 2015 abduction after which my wife had found the two blond hairs, the recall process was sufficient for me to remember most of what had happened with minimal assistance. This means the brain has been successfully trained over time to partially overcome the neurologic manipulation and one can dig into the long-term memory to retrieve most of what had previously been concealed. Practice makes perfect, and a minority of abductees become so successful at this process that they are able to effectively recall 100% of the missing time period on their own. It's not automatic, like short-term memory, but by using self-hypnosis they have taught themselves to get past the memory blocks installed by the abductors.

Interestingly, during this 2015 incident I recognised two people from the same village aboard the UFO. Were either of them aware of their experiences? Did they harbour strange memories, or suspect that all was not quite right in their lives? I subsequently met them both—one worked in a local medical facility—but never mentioned the memory of the encounter. On occasion, an abductee will do just that after they've met people 'aboard' and sometimes will receive a positive response.

The memories continued to emerge regularly for three years or more. When in a deep sleep I would be shaken awake by my wife, who told me that I was "squealing in distress again." Invariably, memories of an abduction would return when the mind was relaxed, and would be to a greater or lesser degree distressing. I then understood why David cautioned abductees about this process in advance. He was absolutely right: once the door is opened and you begin to access the deep recesses of the long-term memory, the only way to turn off the tap is to stop the sessions and walk away temporarily from the subject, which is what I decided to do for a time because the process was becoming so disruptive to my professional working life.

This decision to take a break from the process doubtless closed off the increasingly effortless recall and meant that when I re-started, the process of opening the memory vault had to begin again. David was as patient, supportive, and understanding through this process as always.

The End of an Era

These two researchers, Hopkins and Jacobs, are no longer engaged in the research and have not been replaced by others as competent, grounded, or as intellectually brilliant. Many investigators/researchers are working with abductees, but the revolutionary insights resulting from the diligent, patient work of Budd Hopkins and David Jacobs are now largely absent from the field. Leo (Ronald) Sprinkle, and John Mack are gone; Karla Turner and Barbara Bartholic are gone; and Raymond Fowler, born November 1933, is nine years older than Dave Jacobs. Where are the young, diligent researchers to take up the baton and break new ground in our understanding?

It was of invaluable help to me personally to have worked with David Jacobs, and I shall always feel privileged to have known him. Without his patient guidance I may have never been able to make sense of this unwelcome intrusion into my life.

Retired, and now a widower, David relocated to live closer to his youngest son in the Chicago area. But before leaving Philadelphia he arranged to have all of his work, and much of Budd Hopkins', archived with the American Philosophical Society,[12] which is based in Philadelphia. This includes the recorded sessions of almost all 150 abductees, made available for future reference, with all names and other identifying information carefully redacted (the abductees themselves each have unredacted copies of all their sessions). His library of thousands of books and periodicals on the UFO subject collected over fifty years were all donated to Temple University Library,[13] where they may now be seen and accessed. From 2021 he has a new website, with a fresh look and perspective.[14]

In the next chapter we'll try to answer some basic questions you may have about the program and about the abductees who, wittingly or unwittingly, find their lives inextricably caught up in it.

Chapter Eight

Question Time

*Is the Program really real?
How long has it been going on?
– and other questions*

In this chapter, we shall attempt to address some specific questions about the abduction phenomenon. Some of the answers are easy; others are more complex and nuanced, and some demand that more speculative ideas be explored and entertained.

Is Alien Abduction Real? Why Doesn't Everyone Know About It?

If you were sufficiently interested in the subject to pick up this book and have read this far, it may seem absurd to spend time addressing this question. However, I am mindful that for the non-abductees in the human population—likely to be between 95% and 98% of all people on Earth—the question is a valid one. It's also a valid question for most abductees, who remain unaware of what is going on.

If you have never studied this subject and are unfamiliar with the quirky but identical granular detail in thousands of abduction accounts, or the compelling but elusive medical evidence of near-identical bodily scarring and small bodily implants, then room for scepticism even among fair minded people naturally remains about whether this phenomenon exists at all, or if some other explanation may be concocted to force-fit the facts in service to the 'Felix Frankfurter' mind-set ("I don't doubt the evidence, or that the witnesses are truthful, *but I simply cannot believe it.*")

The facts are as follows:

1. A large number of abductions are reported by people fully awake and driving automobiles or other vehicles, or engaged in other daily activities

2. People are often abducted in groups of two (like the Hills and the Pascagoula pair), three or four (like the Allagash Four) or even more, and their mutual accounts of the experience always match

3. Witnesses watch people being abducted. The Travis Walton case in November 1975 is the classic example of an abduction witnessed by multiple people, but there are others equally as compelling: the 1989 Brooklyn Bridge abduction, although it occurred in the early hours of the morning, was witnessed by more than 20 different people, none of whom personally knew the abductee, or each other

4. People are physically missing for the duration of their abductions (Travis Walton was missing for five days)

5. With a very small number of exceptions, abductees do not welcome what is happening to them and shun any public exposure because they fear they will not be believed, or even that they might be ridiculed. They all want the abductions to stop and wish their lives were normal

6. 'Return mistakes' include abductees being left at a different location to that from which they were taken. Some have found themselves locked out of their houses at night or are returned to the wrong room or wrong bed in the house. One woman was returned to the middle of a wooded area in the dark, a mile from her home, barefoot and in her nightdress; she finally got home with frozen, cut and bleeding feet and then had to wake other family members to unlock the door and let her in. An event such as this might finally convince you that you're having abduction experiences and lead you to seek help, as happened with this unfortunate woman

7. Most compelling of all, the accounts of abductees match in essential detail: the description of the appearance and behaviour of the abducting entities, the nature of communications from the same entities, the procedures to which abductees are subjected, and the interior environment and appearance of the enclosure in which these procedures take place. Thousands of different people can't

make this stuff up: if it were some fantasy, there would be incomparably more variation in the narratives

8. The psychological trauma experienced is compatible with that of genuine trauma from real experience, and not from some internally generated fantasy or delusion [1]

9. Abductees suffer repeat and often frequent experiences throughout their lives

10. They often exhibit identical scars, bruises, burns, biopsy marks, and other signs of physical damage

11. Small implants have frequently been discovered at various locations within the bodies of abductees, and occasionally surgically removed and analysed. The analyses reveal manufactured nano-devices emitting radio waves in specific frequency bands, and the structures are reportedly composed of mineral isotopes not found anywhere on Earth

12. The children of abductees frequently report being abducted, as do their children and their children, down the generations. Why the abductors do this is not well understood. (I propose some ideas for consideration later in this chapter.) But the abduction program is intentional and purposeful: these entities do not prosecute their campaign of exploitation of the human race in a haphazard or random manner. There must be a good reason—or multiple reasons—for this intergenerational program, known only to the abductors. And they ain't tellin'

Any attempt to explain the abduction phenomenon that fails to address all the above factors is, at best, misguided or inadequate; at worst, it's plain wrong and intentionally deceitful.

Sometimes these 'explanations' originate in the personal ideology of a True Believer. Fundamentally you can never argue against an ideology, i.e. "*I just cannot believe* that alien abductions as reported can be happening, so it must be something else these people are going through," is an ideological position rooted in a personal belief system: the ideologue will always shift position to accommodate the ideology because they *just know* for certain what 'the truth' is. There is a saying attributed to Mark Twain: "Never argue with stupid people. They will drag you down to their level and beat you with experience," which needs to be borne in mind here.

Alternately, what is sometimes claimed to be a "rational scientific explanation" is, in fact, a deliberate attempt to deceive, to minimise, or discredit the reality of the phenomenon.

David Jacobs writes succinctly on the subject and with considerable insight:

> From the beginning of the abduction phenomenon debunkers, critics, and proponents have organized their knowledge about abductions based on incomplete evidence and culturally determined attitudes. As knowledge grows, theories must be revised. As we learn more, the verities of the past become the naivetes of the present. All knowledge is subject to change as new evidence is developed. With this in mind, we must revise some of our assumptions in the light of new, sometimes disconfirmatory and even disturbing information. Although there is much that needs to be rethought, I would like to discuss some theories and thinking that presently need critical re-evaluation.[2]

Jacobs continues:

> All UFO researchers are aware of the muddled and "shoot-from-the-hip" thinking that non-UFO researchers, skeptics, and, especially, debunkers have employed over the years. They have linked the abduction phenomenon to a myriad of internally generated phenomena with a wide range of causative factors.
>
> These explanatory systems are based on either a fundamental lack of knowledge of the abduction phenomenon or on a systematic disregard of the disconfirmatory evidence within it. For particularly ill-informed skeptics, the list to choose from is a long one: psychosis, fabrication, fugue state, science fiction, media contamination, folklore, mass hysteria, hypnosis problems, fantasy prone personalities, suggestive people, sleep paralysis, and so on. The more sophisticated skeptics employ these explanations, or a combination of them, and constantly develop new ones when the older ones are shown to be untenable. Screen memories of sexual abuse came into currency in the mid-1980s; temporal lobe lability also of the mid-1980s; fantasy prone personalities, current in the late-1980s; sleep paralysis of the early 1990s; false memory syndrome of the mid-1990s; and by the late 1990s the trendy

"millennialism." These explanations tend to fall out of favor and then make comebacks from time to time, as debunkers refuse to let go of the "oldies but goodies".

As disparate as the critics' explanations are, they have an important commonality: they come in successive fads, one after another. For decades, debunkers and skeptics have attempted to link abduction accounts to causative factors within the society. Debunking explanations tend to be dynamic, changing, and linked to these cultural currents … For example, when sexual abuse became prominent in the press, debunkers seized upon it as an answer to abductions. When False Memory Syndrome became a "hot" problem, its exponents thought they had found the answer to abductions. The same has been true of wave after wave of explanations.

Contrary to the debunkers' opinions however, like the UFO phenomenon itself, the basics of the abduction phenomenon have over the years remained the same regardless of what is current in the society. The essential parts of Barney and Betty Hill's early 1960s account are operative today. Even the essential parts of the 1957 Antonio Villas Boas case are informative and operative now. Rather than the abductions, it is the explanations that have proven to be societally linked and temporally bound. They are, in essence, faddist. In this way, the explanations reveal more about popular culture and the society from which they spring than they do about the etiology of the abduction phenomenon. They reveal a fundamental lack of knowledge and engagement with the phenomenon on a primary level that nearly all debunkers share. And no one has ever been able to duplicate the complex abduction narrative within a clinical or laboratory situation from a person who was not an abductee.[3]

In Ralph Blumenthal's excellent 2021 biography of John Mack, *The Believer: Alien Encounters, Hard Science, and the Passion of John Mack*, the author reminds us of Mack's early work with abductees in his clinical practice at Harvard, and the dawning conviction that he was confronted by a new phenomenon in psychiatry which, contrary to most ill-informed commentators, exhibited all the markers of real experience.

Mack gave a presentation at the 1992 MIT conference on abductions during which he, as a clinical psychiatrist, laid out why he did not think it even possible that the abduction phenomenon was psychiatric in origin. Any explanation, he said, needed to account for five elements:

1. The consistency of the reports
2. Physical signs like scars and witness-backed reports of actual absence for a time

3. Accounts from children too young for delusional psychiatric syndromes
4. An association with witnessed UFOs
5. The lack of any consistent psychopathy among abductees [4]

Furthermore, Mack described abductees when interviewed as exhibiting the *"appropriate self-doubt"* which inevitably accompanies a real experience which is recognisably outside what abductees themselves, and society as a whole, considers to be real or even possible.

An essential feature of the abduction program is that it is, and has always been, clandestine. The entities cannot allow the target population to work out what they are doing, so take great care to ensure that the abductions are neither observed, nor recorded, nor remembered. This secrecy is a kind of force multiplier in human society which assists the continuance of the program in a number of ways:

1. It is easier for debunkers masquerading as 'the voice of expert, scientific reason' to command the stage whenever the subject is discussed. In a world of widespread indifference it is easy to remain wilfully ignorant of the evidence, or to ignore or distort the evidence when presented

2. There is nowhere to go, no channel to follow or official body to report to, if you are having abduction experiences. Effectively, you're on your own until you can link up with others whom you can trust

3. The issue may be more easily consigned to the realm of populist entertainment, which occupies a cherished place in global societies in our current media-obsessed age. To many high-achieving people this helps to invalidate the subject, removing it from serious consideration

4. It is easy for those in high-status positions of public trust, such as members of the scientific community, senior officers in the armed forces, those in higher education establishments, occupying elected office at every level of government, or those in the security services to overlook, disregard, or outright dismiss this potentially very real

threat to humanity because the evidence is not known or not understood, and they simply dismiss it out of hand and refuse to deal with it

The barriers to understanding are considerable. Acknowledging even *the possibility* that these things might be real means that you will need to:

- Interest yourself in the subject sufficiently to devote at least a modest amount of time to examining the evidence. Most people live busy lives, have pressing family, financial, and other concerns and will not put such an esoteric and elusive subject high on their list of priorities

- Get past the depressing reality that the subject has become debased in popular culture to the extent that it is now 'background noise', consigned to the fringe of late-night populist light entertainment

You may remain unconvinced and do a repeated 'Felix Frankfurter' on yourself: an understandable reaction, given the paradigm shift necessary to acknowledge that this just might really be happening. Or you might be honest with yourself and admit: "There's something going on with those reports of abductions, but I don't have the time to look into it all because my life is too caught up with this or that, and I'm not that interested in it because it doesn't affect me." In other words, settle for a life as one of the *Incurious*. Nothing essentially wrong with that, so long as you know that's what you're doing and the reasons for your choice.

When Did This Thing Start? How Do You Know?

Some writers proclaim that the abduction phenomenon can be dismissed with tired pop cultural references, as David Jacobs illustrates in his piece above. They are unaware of the evidence, or else the evidence they know is ignored or distorted in order to maintain their erroneous position and endow it with some thin veneer of credibility.

Others acknowledge reports of abductions to be real but proclaim that the phenomenon is ancient, even as old as recorded history, and point to European and African folklore tales of faeries and other such supernatural entities abducting humans and exchanging babies with human mothers, etc. The writings of John Keel and Jacques Vallée from the 1960s to 1970s made the case that the totality of the UFO phenomenon corresponds to this model, and there is "nothing new here." Though an interesting diversion and ultimately self-indulgent mental exercise, the evidence for these ideas when examined is tenuous, and the *Magonia* model simply does not fit the complex but resilient facts of the abduction program.

The evidence for the abduction program indisputably points to an extraterrestrial origin and furthermore, that this phenomenon dates from the late 1800s, not later and not earlier.

Let's look into that, shall we?

You may recall reading in Chapter One that, in 1966 when I was ten years old, I asked one of the abductors with respectful curiosity (telepathically, because that's how they communicate with us and with each other) why they were repeatedly doing this to me, and why they had singled me out as one of their 'chosen ones'. The taller grey alien casually replied that it was "because of your mother and grandmother *and great-grandmother.*" Full stop. As I was next in the family line they *had* to do these procedures, or had some kind of 'right' or need to do them. Images of these three specific female ancestors when young adults were placed in my mind. Although I had known my grandmother only as middle-aged, and great-grandmother only as an elderly lady who died in January 1963 at age 90, I knew for certain that the images of these young women placed in my mind were genuinely of them.

This was an uncommon example of one of the taller greys voluntarily communicating truthful, factual information to me, but as he told me on a different occasion (perhaps numerous occasions but my clear memory is of only one such), "It doesn't matter what we tell you here, because you're not going to remember anything anyway."

The important thing here is that the alien did not either state or imply that the follow-the-family-line program went back any further than my great-grandmother, who was born in January 1873.

Now, I hear you saying: "But that is what you claim an alien told you during an abduction when you were ten years old, in 1966. That's pretty weak evidence," and I would wholeheartedly agree with you. It's 'anecdotal'. I offer this only as personal experience, something that I remember very well; I could make no sense of it at the time, but there it is.

But there are other pointers.

Abduction researchers over the years have often interviewed abductees about the remembered experiences of their forebears i.e. my grandmother told me about the "pixies" which repeatedly abducted her throughout her life. She was born in February 1908 and, as far as I can see, was describing the 'small greys' from her perspective as a relatively unsophisticated tenant farmer's daughter in West Yorkshire with no higher education and no knowledge of nor interest in the UFO subject, even in the 1960s, so the abducting greys were interpreted as "pixies."

These anecdotal stories of abductions date back to around the 1890s and are not reported *in specific detail* prior to that, though folk tales of faeries covertly exchanging babies and small children with their own kind do exist in European lore through the centuries, which serves to confuse the issue somewhat and gave my grandmother something culturally recognisable to which she could attach her abduction experiences. Of course, the abductions may have been going on prior to 1890 but few reports survive which might reflect the standard abduction narrative, either as oral or written testimony.

The abductions are clearly the cornerstone of a planned and goal-directed program. Its intergenerational feature means that either some or all the children of abductees will also be in the program, whether or not the spouse of the mother or father is involved. With the relentless population increase since the late nineteenth century, the abductee population has spread out into the overall human population in a cone-shaped graph. Dependant on the size of initial population selected, it is possible to calculate the number of generations needed for the entire population to be abductees. According to population statistics, if two percent of the population were originally selected, then after seven generations all humans born on Earth would have at least one single ancestor seven generations back who was included in that original two percent. (Two percent of a population is 1 individual in 50, and you have 128 direct ancestors going back seven generations). Reducing the initial selection number by half, so that just one percent of the original population were selected, we still arrive at an abductee population that is significantly large after seven generations.

It is simply not possible that the abduction program as we know and see it was taking place before the late eighteen hundreds. If that were the case, a much larger proportion of our current population would be abductees.

The estimated total global population of humans in the year 1700 was 600 million; in 1800, 813 million; and by 1900, 1,550 million (1.55 billion).[5] During the 1800s the global population really grew apace, driven predominantly by the availability of clean drinking water; sophisticated sewerage disposal systems which for the first time in history made larger urban population centres relatively free of endemic bacterial disease; and medical advances—like Edward Jenner's smallpox vaccine—that began to attenuate epidemic pathological diseases which prior to that date had for centuries regularly devastated human populations.

Together with these dramatic advances in medicine and public hygiene was the industrial revolution, initially seen in Europe and North America, but rapidly spreading everywhere on Earth throughout the following century. It may have been a combination of these developments which motivated the entities to initiate their program—to what eventual result, we can only speculate. But it looks like they're here to stay.

What were the original selection criteria? No one knows, and they ain't tellin'. I have spoken to abductees who report that human hybrids/hubrids, who can be far more communicative with abductees than the greys (but who are not necessarily more informed), tell them that originally a "virus" was seeded into the atmosphere, and those who "responded in a certain way" were selected for inclusion in the program. Now this source is admittedly about as anecdotal and non-scientific as we can imagine, but these abductees are being told that originally *some* criterion was used. Also, what we understand medically by a "virus" might not correspond too closely to what is being described here. I was told by a communicative hubrid in 2016 that the forthcoming "change" which they talk about as being "wonderful," etc., would be announced by the spreading of a virus in the atmosphere to which abductees and non-abductees will respond differently, and that after that event, "everything will change." This mental download was delivered with striking visual images. But that, too, is just anecdotal. Knowing the abductors' more than occasional tendency to deliberately deceive (= lie outright), you would be wise to remain highly sceptical about any such prophecies from any quarter, including from me.

It may be that the original selection was not particularly important to them, only that they establish—and maintain control of—a series lineage. In the absence of any evidence to suggest otherwise, perhaps it is safe to assume that these *origin abductees* were chosen more or less at random or opportunistically, based approximately on population numbers in different parts of the world. This makes sense even if it eventually proves to be incorrect, but we may never know the answer to this because the abductors don't tell us, and those specific entities who deal with us personally during our abductions may not even know.

We shall return to this question in the next chapter, to examine the issue from a different perspective and look at some surprising evidence from the early stages of the program in the 1890s.

How Many Abductees Are There?

Some abductees have posed this question to the abductors, and if the question is answered at all, it's usually something along the lines of "it's a very large number." I asked one of the taller greys this question and received no response. I tried quoting some numbers, and persisted with "Is it anything like one in twenty?" He eventually responded *"Something like that."* When I pressed the point, he defaulted to the usual *"It's a very large number."* But this particular entity, in his specific role in the abduction program, may not even know the answer to this question. Indeed, he may be possessed of no curiosity at all about the issue, because it's not in his 'job description'.

Now, I know these beings often lie to gain co-operation and calm hostility in abductees, but they do sometimes tell the truth as well, especially if there will be no risk in giving the information because *"it doesn't matter what we tell you, because you're not going to remember anyway."* And they were not lying when they told me that my abductions were *"because of your mother and grandmother and great grandmother."* So, maybe it is one in twenty people who are in the program. Even if not, I think that we can assume that it's *"a very large number."*

During 1991, three separate polls were commissioned together by Budd Hopkins, David Jacobs, and sociologist Dr Ron Westrum, which were funded by Robert Bigelow and carried out by the Roper Polling Organization. They sought to determine a rough estimate of how many Americans might have experienced phenomena common to the abduction program. The polls were conducted at people's homes in personal interviews carried out from July through September of that year.

Mixed in with questions about political beliefs and voting intentions, the economy and ecology issues, 5,947 Americans were asked early in the survey whether they had personally seen a ghost, or ever had a UFO sighting. There was also a control question: whether the word 'Trondant'—a made-up word—had any particular significance for the respondent; the one percent of respondents who answered positively were excluded from the results, as it was judged that their responses could not be relied on. (Pollsters agree that one percent of respondents will *always* answer a question affirmatively, regardless of the question asked.)

Respondents were then asked whether they had ever experienced any of the following phenomena. The surveyors were unaware of the connection, nor of the reasons these questions appeared on the survey. The five indicator questions were as follows.

Have you ever:

- awakened paralyzed, sensing a strange figure or presence in the room?

- experienced an hour or more of 'missing time'?

- felt like you were actually flying through the air without knowing why or how?

- seen unusual lights or balls of light in a room, without understanding what caused them?

- discovered puzzling scars on your body, without knowing how or why they got there?

If people responded positively to two out of five of these experiences, it was reasoned there was a good possibility they might be UFO abductees. The results suggested that as much as five percent of the population, or one person in twenty, were probably abductees. These numbers were considered to be much too high, and were rejected, if only to avoid having the debunkers and 'skeptics' charge that the poll was in some way rigged, or that the criteria were much too liberal. The criteria were therefore tightened up so that only those answering four out of five questions positively were accepted as probable abductees.

By these much stricter criteria, of the nearly 6,000 Americans polled, 119 answered in a way which suggested they might be affected by this phenomenon.

In other words, that they were probably abductees. From these figures, the pollsters concluded that as many as four million Americans have endured the abduction experience—almost two percent of the population.[6] "So, you see, it is not a rare phenomenon," proclaimed Dr John Mack, when presented with the results.

But the results using the poll's original criteria revealed five percent of the population may be abductees. The pollsters were confident that the two percent result from the 4/5 group was a near-certainty.

This exercise has never been repeated but probably ought to be in the 2020s, as the original poll data is 30 years old. If the poll data in 1991 may be used as a rough guide, then two percent of the 2022 global human population of 7,948,118,521 would be 158,962,370, or almost 159 million people. Five percent of the current global population of 7,948,118,521 is 397,405,926, or almost four hundred million abductees in the global population.[7] Anecdotally, I have met with dozens of abductees, many who are seeking answers or guidance as to what to do, where to go, and who might assist them with recovering buried memories. These are only the aware abductees, so the tip of the iceberg above the waterline, and exclude the vast majority of the unaware. I think we can confidently say that it's "a very large number." This thing is obviously very widespread and affects millions of people.

As the aliens tell us, it's a big number, and the implications are that it's a big program. What they are engaged in is obviously of great importance to them: the resource commitment must be enormous, and it's been prosecuted twenty-four hours a day, relentlessly and all over the Earth, for more than a hundred years.

How Do Some Abductees Know, and Others Not?

The abduction program is *clandestine*. The entities take great care to ensure that the abductions are neither observed, nor recorded, nor remembered. This means that abductees, once they initially become aware, and eventually certain, about what has been happening to them, are forced to make great efforts to remember or find assistance in overcoming the memory blocks. This is at the least challenging and difficult, and well-nigh impossible for many people. Whether it's an inability to travel, slim financial resources, or even a lack of personal confidence to approach anyone 'in authority' with whom one might share what, understandably, is difficult for many to believe, the obstacles that an abductee faces when first attempting to come to terms with this may seem insurmountable.

The vast majority of abductees, however, remain unaware that abductions are taking place at all, not recognising that this phenomenon might be the reason their lives are so strange in so many ways. Many, even when confronted with credible information about the subject, simply do not consider themselves 'one of those people' and remain convinced that this thing, though perhaps interesting, *absolutely does not* apply to them. With the exception of a very small number of deluded narcissists possessed of some New Age worldview, nobody wants to be an abductee.

Abductees often report a 'trigger event' at a certain moment in their lives which breaks through and convinces them that something extraordinary is happening. This event carries the certainty that it cannot be denied or ignored, regardless of how uncomfortable it makes them feel or how unwelcome the knowledge of such an intrusion into their lives proves to be. If it is recognised as an abduction event they will conclude with astonishment, "I was abducted by aliens when I was twelve!" (Or at whatever age this incident occurred and was remembered.) But they will invariably view this as a one-off event, lacking conscious memory of other incidents or knowledge of how the phenomenon works. Despite such a fantastic realisation, they will not extend it to conclude that they have been lifelong abductees, not yet grasping the lengthy pattern of similar experiences. The abductors have set up the program in such a way that abductees remember little to nothing, and the traces they leave—the bodily scarring, the small implants and so forth—may be ignored or overlooked by the majority of abductees.

It is difficult in this age of information saturation to avoid completely the meme of alien abduction in popular culture, but extremely common to have little understanding of what it really is all about. It requires determination to seek out reliable sources of information on the subject and to ignore or avoid the irrelevant, the misguided, and the intentionally misleading chaff. It's no wonder then that the majority of abductees in the global population have no idea what is going on and do not even suspect they themselves may be a serial abductee.

As such, the program has been very successful. It has been unwittingly aided and abetted by the human population, its cultural norms and paradigms, and greatly assisted by the ubiquity of popular culture and the silent neglect of the scientific community, not to mention those who have chosen a career path dedicated to the protection of the population yet who continue to ignore it. It's not surprising that the vast majority of abductees remain completely unaware that this phenomenon is intruding into their lives and lurks beneath the surface of normal consciousness.

The 'Intergenerational Component': Why Would They Do That?

This is a hard one to answer because the truth is that nobody really knows for sure. The only certainty is that they *do* follow genetic family lines, and *do not* abduct those who are born with neither parent in the program. The very occasional exceptions to this rule are when a non-abductee is caught up in the abduction of one or more regular abductees and, rather than being 'shut down'—kept in an immobilised state until the conclusion of the event—is instead taken with them. Chuck Rak and Barney Hill may be examples of this. It's unlikely that either of them were ever abducted a second time. Rak recalled under regressive hypnosis, ten years after the Allagash event, that he spent most of the time inside the UFO immobilised, ignored by the abductors and watching Charlie Foltz undergoing medical examination procedures.

So, if you're a regular abductee, it's almost certain—so far as we know—that you're descended from an origin abductee, probably chosen sometime during the 1890s.

There are occasions when paternity is not always easy to determine, or other related problems arise. A non-abductee woman may be away from home and have a brief affair (like a holiday romance, for example), and later finds that she has become pregnant. She decides to have the child anyway. The father could, in fact, be an unaware abductee living far away in another continent. Will the abductors know and recognise the child as one of their own, and will that child be inducted into the program? If so, how would the abducting entities know? It's impossible to gain any data on the frequency of such cases, so we just don't know how often this has happened.

But the limited knowledge I have of the program and the aliens' standard operating procedures just tells me that *the abductors will know, somehow*. There may be genetic markers in abductees after generations, or other ways they can tell—for example, the abductors can read the minds of abductees and know their recent history of activities via a 'mindscan' procedure—but they *will know*, and the child will likely be found and abducted into the program like his/her father.

Human females have been used by the abductors for the gestation of foetuses for periods of around ten weeks, before the foetus is removed and transferred to gestation in vitro inside a tank of fluid. (NB this stage of the program may now be at an end.) Thousands of abductees have reported being shown these gestating foetuses, which grow into hybrid children and then into adults. Many female abductees have in the past reported feeling pregnant when they know they can't be by normal means, confirmed by a positive conventional pregnancy test, and then mysteriously after a few weeks apparently losing the pregnancy but with no evidence of any miscarriage.

The obsessive continuity of the instigators of this program following genetic lines suggests that the process involves incremental genetic modifications. Perhaps they are breeding something into the abductees' genetic code which requires both sperm and ova from the now genetically determined abductee population.

We shall now digress for a while into exploring one of the minor mysteries of the abduction program, which may relate directly to, or even partly explain, the intergenerational aspects of the phenomenon.

A human female embryo starts to manufacture ova nine weeks after conception. By the time of birth, the female child's body has made several million oocytes (foetal egg cells) but by the onset of puberty and ovulation, the vast majority of these oocytes have died, with approximately 700,000 immature ova remaining. During her reproductive years the adult female will release somewhere around 500 mature ova, one approximately each twenty-eight days, excluding periods of pregnancy when ovulation is paused, from the onset of puberty until the menopause. Ova may also be extracted mechanically—egg donation is a legitimate medical procedure—but after birth, *no new ova are ever produced*. So those extracted, or released from ovarian follicles during each ovulation cycle, are the survivors from pre-birth production. Following the menopause, there are no more ova.

Sperm production in human males follows a diametrically opposite pattern. In contrast to ova production in females, sperm cell production is not even initiated until puberty, and new sperm cells are produced by the testes continuously for the remainder of a man's life. Though the rate slows down considerably as the man ages, he's even producing some sperm if he survives beyond one hundred years, though both the quantity and motility are considerably reduced. The point is that these cells are *all newly manufactured by the testes*, whatever the man's age, and this process continues until death.

It has been reported by virtually all male abductees (some don't remember) that the abductors are very insistent about obtaining a sperm sample during each abduction event. Often, the abductors go to great lengths to obtain these samples: if the usual mechanical methods—a kind of suction device placed over the whole groin area while the man is lying immobilised on the examination table—prove insufficient, they will resort to the assistance of either female abductees or female hybrids to get the desired result.[8] They have occasionally even been known to extract sperm by a kind of electroejaculation stimulation (EES). Men who have had vasectomies have reported to me that the abductors have a mechanism for dealing with that too, which involves the insertion of a thin needle-like device directly into the area immediately behind the testes. This extracts sperm cells directly, which excludes the seminal fluid and seminal plasma.

Why not just take a single sperm sample from each male abductee and freeze it? Such a sample might subsequently reap millions of hybrid offspring. Aliens with advanced technologies surely must know about the preservation of biological samples by freezing. From the perspective of the current biological sciences—notably, the practices of IVF clinics—this may be true. But there may be a good reason they continue taking samples, which aligns with current assumptions of what their program is about. It could be that the continuous, subtle genetic modifications are measurable in the sperm, so that a fresh sample is procured each time. The hybridisation-gestation program would thus utilise freshly manufactured sperm, bang up to date with any such subtle genetic modification.

This regular sperm sampling has been reported for decades and confirmed by thousands of abductees, so is obviously of major importance to the abduction program. The hybrid breeding program is central to the abductors' overall agenda, and the intergenerational continuity an essential part of it. They seem to have no use for new abductees, whose direct ancestors have never been in the program, and they can somehow always tell the difference.

What Is Life Like For an Abductee?

The answer to this question is that it largely depends on the character, physical constitution, and circumstances in which an abductee finds him/herself. A minority simply remain determinedly incurious about the extraordinary, have few spiritual or religious inclinations and attach themselves to career, family and leisure interests compatible with popular culture: they remain content and settled (sort of). For tens of thousands of others, it's a different story.

Speaking for myself, I confess that the knowledge that something deep and disturbing was going on 'under the surface' was a permanent feature of life when growing up and well into adult life. I put great efforts into career development and worked very hard at it for years but could never escape the 'other thing' which seemed to lurk beneath it all. Many of the pointers in the list in Chapter Three, titled, *Finding Out if You are One of Their 'Chosen Ones'*, unquestionably apply to me, including those events directly indicative of a recent abduction event.

In May 1991 Michael Lindemann, president of the 2020 Group, conducted a recorded interview with Budd Hopkins. During this encounter, Lindemann asked what kind of people abductees are, if any common attitudes or personality traits are evident in this group or can be seen to emerge over time. Hopkins was an extremely astute observer of human nature, as the excerpts below reveal:

> There are certain things that I think I've learned to recognise in people, and this is totally subjective, probably indefensible. But at any rate, I would say that most abductees that I've worked with have an immediately more open, broader attitude toward any kind of metaphysical, spiritual question or possibility. They are not necessarily people who respect authority automatically, by rote. In a strange way, they have seen a larger universe, closer-up, than most of us have, whether they remember it clearly or not. As one person said, it's as if, especially when they begin exploring their experiences, they've gotten rid of the tunnel vision that most people have suffered from.
>
> On the other hand, I've never met anybody who I think was helped by this. The psychological scars are there for everybody. Often, in a room full of abductees at a support group meeting, I see a lot of extremely attractive, interesting, intelligent people, and realise that there are very, very few successful relationships in that room amongst the people. It's tougher for someone who has been through this. The self-doubt, even a kind of odd shame about their experiences, precludes a kind of easy, relaxed interaction with other people. If you see someone who is interesting, intelligent, open-minded, someone who accepts a wide range of possibility, and yet that person has a lot of trouble, which they shouldn't be having, with relationships and self-esteem and everything else, those are earmarks of having gone through these abduction experiences.
>
> I don't want that to sound depressing for people who've been through these experiences. The degree of success of many, many people I've worked

with is enormous, yet they don't really believe in themselves. One person is a nationally known star in the entertainment business; another person is a self-made millionaire from a poverty-stricken background with a very important place in the real world—yet neither of these people have the self-esteem of Dan Quayle. [Quayle was a notoriously ignorant Republican politician who once, when serving as George HW Bush's VP, insisted on misspelling the word "potatoe" in front of a class of sixth grade children on nationwide television.]

It's unfortunate. There's an enormous disparity between their actual talents and abilities, their actual accomplishments, and the way they view themselves. And I think that's one of the legacies of these experiences.[9]

Hopkins is absolutely right in his assessments: abductees tend to have the diametric opposite to narcissistic personalities. In other words, they are often genuine high achievers, but have little confidence in themselves and their success tends to come from a quiet but dogged determination to be good at something (or at many things) despite not feeling, deep down, really connected or committed to their professions at all. Speaking from experience, something deeper is always pulling at you and you work at something because you are determined to resist the abductors and what they are doing, even if you are consciously unaware of their activities and effects on you.

As Hopkins observes so astutely in the Lindemann interview quoted above, abductees tend to have a *"more open, broader attitude toward any kind of metaphysical, spiritual question or possibility."* Many of them don't actually welcome this, but instead wish they were more 'normal' and not so attracted to the weird and the 'fringe'. Yet there's nothing you can do about it; you are compelled to embrace and look into these things by a powerful, even obsessive, fascination. You always feel like an outsider. It's only when you meet other abductees that you start to make some sense of it, because you recognise that these people don't want to feel the way they do any more than you do. You're kind of stuck with it, and constantly work to accommodate and integrate these two separate parts of your life.

David Jacobs' first book on the abduction issue had a resonant title: *Secret Life*. That's what it's like.

Coming up is this book's final chapter, which examines what we think we know with some confidence about the abductors: their possible origins, intentions, and methodologies; the timescale and possible purposes of the program including how, when and why it may have begun, and where all this might be leading. We also examine why abductees consistently describe encountering specific different types of beings, what these different entities might be, and how all this fits together.

Chapter Nine

"They"

Where do they come from?
Why are different beings described?
What's the endgame?

From the early 1970s, whenever my mother talked about the entities that were interacting with her—and with me—she simply referred to *they* or *them*. *They* do this and that; *they* take items from the house, and sometimes return them, sometimes not; *they* ask about this and that; *they* were here last night; *they* were always around. She was mystified and bemused in equal measure by *them* and their persistent, intrusive presence in her life. She never saw anything sinister or nefarious in their presence or actions; just bemusement and a kind of permanent bafflement.

So who—what—might *they* be? What is the back story, what is their agenda, and what is the planned endgame?

This chapter will attempt to summarise what we know about those responsible for the abduction program. There is little of which we can be certain, but we can be reasonably confident of some things. For example, that there is a deliberate 'program' that has been ongoing across several generations. Or who the 'real' aliens might be. (Spoiler: they are *not* likely to be 'the greys' which many thousands of abductees reported they dealt with almost exclusively until recent years.)

As with the previous chapter, we delve a little into the dubious area of speculation, this time into some metaphysical concepts at the edge of currently accepted scientific knowledge. But let's begin with what we know of the universe and proceed from there.

The Age of the Universe

It is not within the remit of this work to enter into a long or detailed discussion of complex cosmology and astrophysics: the interlocking of, and seeming contradictions between, on the one hand Albert Einstein's General and Special Theories of Relativity, which describe space-time and the nature of gravity and its effects on a cosmic scale; and on the other, the ideas developed by Niels Bohr, Werner Heisenberg, and Erwin Schrödinger about the quantum levels of interaction demonstrable at the atomic level. The published works of Michio Kaku (especially *Parallel Worlds*) and the late Stephen Hawking are where the reader should look to gain a better grasp of these esoteric and often philosophical concepts.

There is widespread consensus in the scientific community that the age of the universe is about 13.8 billion years, give or take (obviously this consensus does not extend to religious fundamentalists of various kinds, whose faith-based views of reality are permanently immune to fact and reason). This calculation was arrived at through the confluence of several proofs. To begin with, the Big Bang theory posited that the universe had abruptly expanded from 'nothing' in the far distant past. This hypothesis argued against a 'steady state' nature to the universe, in which everything remains more or less the same to (and from) eternity. In 1953, George Gamow [1] predicted that, if the Big Bang theory were correct, then there should be a measurable background radiation, in every direction, which was the 'echo' of the beginning of the universe. This cosmic microwave background (CMB) was confirmed by Arno Penzias and Robert Wilson, of Bell Labs, who detected it during the early 1960s as an otherwise inexplicable interference with their precision radio equipment. Penzias and Wilson were awarded the Nobel Prize for their discovery.

The Big Bang theory also predicts that the universe will be expanding at a calculable velocity. The rates of recession of distant galaxies away from us—for the universe is expanding in all directions, regardless of where one is situated within it—can be calculated by examining the *red shift* of the light as it reaches our telescopes because it stretches out toward longer wavelengths the further it travels. The distance to a galaxy is proportional to its recessional velocity, with the constant of proportionality set as the *Hubble Constant*, H, which turns out to be (approximately) the reciprocal of the age of the universe. That is, we measure the age of the universe by measuring these recessional velocities of other galaxies with the formula $T = 1/H$.[2]

T = $1/H$ is only correct if the universe is not significantly accelerating nor decelerating in its expansion rate following the Big Bang. If the rate of expansion is constantly accelerating, the universe may be older than the $1/H$ = 13.8 billion years currently accepted. Such a pace of expansion would be caused by a larger value of the Cosmological Constant, a sort of anti-gravity force predicted by Albert Einstein's Theory of General Relativity. There is some evidence that this might be the case: this 'dark energy' has stimulated a great deal of scientific work in an attempt to understand and characterise it.

Is Anyone Out There?

The point is that the universe is very, very old. Billions of stars in the centre of our galaxy have existed since the galaxy was young, so they too—presumably together with any exoplanets orbiting them—are very, very old. Our Solar System is a relative newcomer, present for about one third of the age of the universe as a whole since the Big Bang. This calculation is based on the radioactive decay of certain isotopes, such as potassium and uranium, in both terrestrial rocks and meteorites. These elements were created with the Solar System, and we can calculate with relatively high accuracy how long it has been since their creation, so the age of the Solar System—including Planet Earth—is generally agreed to be 4.568 billion years.

Paleontologists calculate the Earth has been host to organic life for between 3.5 and 4.1 billion years, with complex organisms such as reptiles, mammals, and birds developing much later. Humans and their ancestors have been here for about 6 million years; *Homo sapiens*, the modern form of humans, for about 300,000 years and the earliest organized human civilizations started forming only around 10,000 years ago.

Comparatively, humanity as a complex agrarian society has existed for an infinitesimally short period—about 0.0000007% of the estimated 13.8 billion years of the existence of the universe. That brief lifespan could be a trait we share with many alien civilizations, or it might be that some have survived and built highly evolved technological civilizations lasting literally millions or even tens of millions of years, which against the estimated 13.8-billion-year age of the universe would still be a very brief period comparable to switching a light quickly on and off in a darkened room on a long winter's night.

There are tens of billions of stars in the central part of the galaxy that are more than three times the age of our Sun. If only a small fraction of these contains one or more exoplanets, then this ancient region of the galaxy might be host to thousands of highly developed and ancient civilizations. As I write this, the total number of stars with their own planetary systems that have been catalogued exceeds 5,000. Not every one of these has been confirmed to contain an Earth-like, rocky planet within the 'temperate zone'—not too hot; not too cold—but the evidence gathered so far suggests that they might be more abundant than not. And it's beginning to look as though the majority of stars do contain some sort of planetary system.

Cosmologists and astrophysicists assume that it is likely that another highly developed co-operative civilisation has arisen on at least one of the myriad exoplanets now understood to exist among the—*at minimum*—one hundred billion stars of just our galaxy alone.[3] Scores, or even hundreds, of civilisations may have arisen then perished, whether through natural cataclysm or one of their own doing. It's worth noting that nearly all life on Earth has been wiped out—on several occasions—long before humans first began scratching out an existence. Some civilisations might have outgrown a single planetary ecology and by now exist independently of any one planet or star system, as 'colonists' throughout the galaxy.

And what if one such highly advanced civilisation were not only to overlap with our own humble timeline, but also find us? What if more than one race is visiting us? Despite the vastness of the galaxy, with its billions of stars, the likelihood of this is also high.[4] Alarmingly so. And the evidence—the persistent UFO phenomenon—is that this has already happened.

The ETH

From the advent of the modern UFO era in 1947, when the term 'flying saucer' was first coined, speculation about their identity has understandably remained high. This fascinating mystery has generated scores of explanations, from the rational to the ridiculous, many of which seemed designed above all to come up with *any* reason that they could not be from another planet.

During the Second World War, the 'foo fighters' were generally believed to be German weapons—a reasonable conclusion in the midst of a conflict which spurred such huge leaps in numerous technologies. But, by the 1950s, many still believed that the saucers were Nazi technology that had been captured by the Soviets. This notion could be shot down for a number of reasons, and yet it would continue to be the pet theory of many who refused to consider that we were being visited from outer space.

Indeed, following the revelation in 2017 about the Pentagon's AATIP research of UAP,[5] some self-described 'experts' on television are still claiming that these objects might be made by the Chinese! How clever of the Russians and Chinese to design and build, then fly these things *in US airspace* for eighty years, without us matching them, despite the trillions of dollars that are poured into defence programs. It was a dumb hypothesis in 1955 and it's even dumber today.

Many of the other explanations are even more foolish: time-travellers from humanity's future; the Hollow Earth theory, in which a highly developed race of superhumans or 'cryptoterrestrials' resides within the planet; beings that reside within other dimensions, or parallel universes; a kind of psycho-social phenomenon, as propounded more than fifty years ago by writers like John Keel and Jacques Vallée; a 'breakaway civilisation' of humans in possession of exotic technologies—*Moon Nazis!*—funded by 'black projects'. There appears to be no limit to the escalating improbabilities that have flooded our information space.

The single cogent explanation for the UFOs remains one of the first to be put forward when they first became a subject of widespread curiosity: the *Extraterrestrial Hypothesis*, or ETH. However fantastic this notion is, it remains the most rational. Yet there are many who resist going 'there', often with objections about technologies which don't—*cannot*—exist, as though the great leaps made by mankind during the 20th century were nothing. "We can't get there, so nobody can."

But there is another reason for this reservation, which is that the 'saucer people' haven't followed the naive script that popular culture has written for them. They haven't 'landed on the White House lawn', for example. They haven't made 'contact' to share with us the secrets of their technologies and humbly lead us into a future of grace and harmony. In short, it is the *lack of any apparent intentionality* which leads many to doubt that we are being visited. Why would they come all the way here and not ... hang out with us?

The idea that aliens may be intimately involved with the human species in a covert operation to physiologically exploit us, and have no intention of making 'contact' for our benefit, does not seem to occur to many people. It conflicts with the mindset that all extraterrestrial visitors must by definition be highly developed spiritually and have benign intentions towards us. It is this unfounded conviction which has led to the kinds of wild speculation mentioned above, silly ideas lacking any substantive evidence put forth by those who cannot fathom that visitors from another star might be somewhat standoffish.

The evidence that the UFOs and the agencies which control them are from another star is overwhelming. The isotopes in the materials from which the implant devices are constructed show clearly that they are non-terrestrial in origin. And the lack of open contact by the aliens is clearly because *they don't want to share their intentions with us, because their plans are not fundamentally in the interests of the human race continuing its own independent evolutionary development as a separate species.*

Where Do They Come From?

Utilising a nine-gigapixel image from the VISTA infrared survey telescope at the European Space Agency's Paranal Observatory, an international team of astronomers in 2012 resolved more than 84 million stars in the central parts of the Milky Way.

This number is but a tiny fraction of one per cent of the estimated number of stars in the Milky Way galaxy as a whole. The number of stars may be estimated by different methods and the answers differ depending on the value chosen to estimate the average mass of a star. The scientific consensus is that the Milky Way contains around 100 billion stars (that's 100,000,000,000) as a minimum and possibly upwards of 400 billion.[6] And that's just our galaxy. The known universe is composed of at least two trillion galaxies (that's 2,000,000,000,000 of them) – and that's only those we've so far detected, each in turn containing hundreds of billions of stars. Our astronomers, as described above, have now identified more than 5,000 different exoplanets, but there may be *more than a billion* that we've not yet found. That's an enormous factor to plug into Frank Drake's famous equation [7] for making a rough determination of the number of extant civilised planets.

That's why it is unlikely that we will soon learn where they come from. The star system of origin—if indeed the origin is only one star system and not many—may not even be catalogued by our astronomers, let alone named. The star system of origin probably hosts at least one exoplanet, in the *Goldilocks Zone* with ambient temperatures in the range of 0–50 degrees C, capable of sustaining carbon-based organic life, which gives us something to look for. But these kinds of star systems may be ubiquitous. Speculating about its location is a futile exercise.

How Did They Get Here?

It's common for astronomers and cosmologists, with their current level of knowledge of relativity and quantum theory, to declare that extraterrestrials "can't get here from there" due to the incomparably vast interstellar distances and the supposed impossibility of traveling faster than the speed of light. Even if "there" were the closest star to us—Proxima Centauri, more than 4 light years, or some 40 *trillion* kilometres away—it is estimated that the journey would take more than 80,000 years using our best technology. That's time enough for about 2,700 human generations. With our current common understanding of physics that may be true, though new ideas are challenging these old paradigms: there are many theories circulating among leading physicists that faster-than-light travel should be possible.

A more appropriate question is: *Are they here, or not?* If the answer is, "Yes, the evidence shows that they are here," then how they got here is merely an engineering problem. (A non-trivial one, to be sure.) It's only impossible with *our current understanding* of physics and cosmology, which is unquestionably incomplete. If they are here, they obviously got here somehow. The questions therefore ought to be:

1. What are they doing here? (Or, as President Truman allegedly put it: "What do the sons of bitches want?")
2. Why don't they communicate with us in the open manner that our science fiction writers have taught us to expect?
3. What evidence exists of their objectives and intentions?
4. How do we deal with their presence, and what attitude should we adopt to these entities clearly in possession of exotic futuristic technologies?
5. Are they a threat to our continued existence as masters of this planet?
6. Is there anything we're able to do about it? Do we *want* to do anything about it, or should we just accept what's coming?

Since the earliest times when the UFO phenomenon was recognised as a reality and probably extraterrestrial in origin, a minority of humans all over the world have claimed to have had contact with the occupants, or, more usually, claim the occupants/crews of these craft have made personal contact with them. As we have explored in previous chapters, some of the most compelling reports have been made by abductees: compelling because, above all, we have a large number of near-identical reports from a wide variety of individuals, from a wide variety of locations. Let us therefore summarise what we know of these entities from the witnesses' descriptions.

The Different Kinds of Entities Reported

Various distinct types of entity have been reported by abductees. Some researchers believe that, despite the range of different beings reported, they all seem to be devoted to the program, and even to be working together. What follows here is a summary from fellow abductees and researchers, and from my many discussions with David Jacobs, whose work has shed more light on this important subject for me than anyone else.

The Greys

For many decades abductees predominantly described being taken by, and interacting with, the two distinct types of 'greys': the shorter kind, and the taller. The spindly grey humanoid alien with enlarged cranium and huge dark almond-shaped eyes is something of an icon in popular culture, among UFO enthusiasts and disbelievers alike.[8]

These entities have for years been accepted as "the aliens." Consensus reality does not include little grey aliens outside of fiction: had these things not happened to me, then I surely would have enjoyed the luxury of choosing not to believe it either. But *I have seen them*, in the flesh, in a hotel room in Sardinia: they stood just a few feet away from me. I remember that clearly—did not require hypnosis to recover the memory—and would testify under oath as to their existence, and exactly what they look like. Possibly millions more abductees can testify to having seen these beings close-up and personal. They exist, all right: the descriptions of all these witnesses match up and they are not a construct of the human imagination.

They seem to be dedicated to performing specific tasks and to do nothing else. In fact, some abductees see them—particularly the smaller ones—as "like robots" in the way that they perform these functions, as though they'd been designed and constructed for the purpose. This 'robot' proposition is not a view I personally share: as far as I can tell they are sentient, living biological beings with an albeit limited emotional framework and the smaller type do feel fear, anxiety, etc. if they feel their personal safety might be threatened, so like most living creatures, they possess evident strong instincts of self-preservation. The taller ones in particular can sometimes express satisfaction about their work (they seem for example to be proud of their success with the breeding program and the rearing of hybrid children), but this never extends to joy or exuberance, nor indeed to any strong emotion.

There is no banter or chit-chat among these beings as, for example, would be common between staff in a hospital operating theatre environment in the human world. Some abductees occasionally report seeing them "sleeping" or "re-charging" in a specific area (a "room") inside the craft, and there is some suggestion that they regularly absorb nutrients through the skin, because they don't seem to eat: their mouths have never once been reported to open, and there appears to be no alimentary canal or digestive tract. Moreover, there are no reports of either the vestigial mouth or nose being in any way functional. The torso is as thin and straight as a tree sapling, with no evident internal capacity for the multiple internal organs in the human abdomen, such as the lungs. The stick-like chest is never seen to move as with terrestrial mammals when air is inhaled and exhaled. Everything is stripped down to the essentials necessary to perform their allocated tasks with the minimum of physiological maintenance required.

In summary, these beings, so far as reported:

1. Are bipedal humanoid with a large head at the top containing a pair of huge, dark almond-shaped eyes; a hole on either side of the head where human ears would be; a vestigial mouth and small nose which seem to not perform any of the functions that a human mouth and nose does: breathing, eating/ingesting, speech, taste and smell all seem to be absent [9]

2. Display no evidence of possessing any lungs, trachea or respiratory apparatus, digestive system, vocal cords, reproductive organs or secondary sexual characteristics, though the taller kind are identified by abductees as 'male' or 'female' by their nature, bearing, and attitude, and to some extent by their roles in the program. The 'male' ones perform most—not all—of the surgical procedures and the 'female' ones predominantly look after the hybrid children. Both types engage in 'mindscan' procedures

3. Have outsized heads for their bodies, presumably housing a brain with considerable powers of control over the neurological systems of humans, from a distance of at least tens of metres and perhaps more

4. Seem to have underdeveloped muscular systems with lightweight, skinny bodies capable of moving them about and keeping the head working, but not designed for athletic exertion

5. Communicate telepathically with abductees and between themselves but rarely say much or yield up information when asked, other than palliatives such as, "You'll be OK" and, "We're not going to hurt you." Or occasionally the more puzzling, "We have to do this" or, "We have the right to do this to you", or "You know what we are doing"

6. Have long arms and fingers which are used principally to examine and manipulate human abductees during the procedures carried out on them, and to deploy tools and surgical instruments obviously designed and produced specifically for use on the human anatomy. Some reports are of three fingers with opposable thumb on each hand, but occasionally four fingers plus thumb are reported.

What they do is *work in the abduction program*. That is *all they do, all the time*.

Tall Mantis-Like Beings

These tall, slender beings, almost invariably described as "like praying mantises," have occasionally been reported to appear with both the short and tall greys. Whenever one is present, though, there is almost always just that single one. They are described as the tallest beings encountered by abductees, standing between seven and eight feet tall and with a forward-leaning posture.

Though described as insect-like, these beings have only four visible limbs—two arms, two legs like a humanoid—rather than six. They have a triangular-shaped head dominated by large, eliptical eyes; a long neck; a very lean torso and long, lean limbs. They have quick movements to match their appearance.

Occasionally one is described as wearing a long robe or cape "with a high collar." It seems improbable, but that specific point has been reported by a number of abductees. Perhaps this is a badge of rank in the hierarchy. Indeed, they have occasionally been reported to address groups of greys, hybrids, and abductees together and demonstrate some authority and control over whatever is going on around them.

They are almost certainly the instigators of the program. They have apparently built a technologically complex society far beyond ours and the species is almost certainly much older and more developed in a number of ways. They are clearly in charge and seem to be highly intelligent, even becoming impatient with the greys at times. They will occasionally perform the 'mindscan' procedures. They often will interact with an abductee to a degree that is usually lacking during these events, exhibiting enthusiasm about the program's execution, progress, and objectives. Their communication can even be said to verge on loquaciousness compared to the greys.

These communications are all telepathic. They seem able to transfer complex information directly from mind to mind: abductees occasionally report 'instant downloads' of large quantities of information from one of these beings following even a short interaction, which the human recipient then may take hours or days to process. They have complete neurologic control over less developed species, pacifying or paralysing them at will. They can cause us to think and see things and may implant false memories, which the abductee then believes are real memories of incidents which really happened to them. They are complete masters of their environment. It's an open question whether these abilities are the result of natural evolution over millions of years, or were enhanced by genetic engineering. Given the focus of their activities here on Earth, the latter seems quite probable—that at some point they took direct control of their own development.

There is a remarkable artistic rendition by artist Stephan Smith of one of these beings on p.61 of Kathleen Marden's and Denise Stoner's co-written 'The Alien Abduction Files' (see the Bibliography).

The Human Hybrids and 'Hubrids'

Hybrid beings, looking like a cross between greys and humans, have been reported for decades. They have been described as foetuses gestating in tanks of fluid, as well as children and adults. Debbie Jordan-Kauble and others first described these beings *ca.* 1983–84, and hundreds of abductees have since likewise reported seeing them. Clearly, the ubiquitous sperm- and ova-harvesting procedures; the implantation, and later extraction, of foetuses from female abductees after around ten weeks; the reported *in vitro* gestation in tanks of what is presumed to be nutrient-filled fluid—all of this points to a long-term 'hybridisation' program, which we can speculate may be their primary purpose on Earth, or else be an essential stage in service to that ultimate purpose.

Eventually, beings that looked almost human began to be widely reported. Frequently, these beings would take an active part in the abduction process, even taking sole responsibility for the procedures. It has often been reported that they seem to work at acting like us, as though anxious to blend in among humans, even to the point of wearing wigs.

And, indeed, during the past twenty years a new category of hybrid beings—what Jacobs refers to as 'hubrids'—began to be reported. They seem purposed to integrate into human societies, and already live here together in small groups for security. Like the others, they can control the minds of abductees. Although this control is to a weaker extent than that of the greys, they can still effectively control any human they are with, and more easily when a group of them act together. They are naturally telepathic but also learn to speak orally in human languages. Abductees' interactions with these beings are described in detail in Jacobs' third book on the abduction phenomenon, *Walking Among Us*, and widely reported from other sources, including in Hopkins' and Rainey's *Sight Unseen*.[10]

This is a specific stage in the process of what Jacobs refers to as *planetary acquisition*, which he reluctantly admits is what appears to be the goal. I shall not explore this aspect of the program in detail, but instead direct curious and interested readers to read Jacobs' excellent work on this whole subject.

Reptilians

Encounters with reptilian-looking creatures have been consistently reported over the years, and not only by abductees. These beings are described as tall and imposing, with lizard-like (usually) yellow eyes with vertical black slits for pupils. They are variously described as "like a crocodile," or "like an iguana," or as having snake-like heads. The wide variation in their appearance is especially intriguing. Such beings have appeared in religious art throughout the ages, particularly in oriental cultures. Widely considered to be 'demons', they are rarely seen as benign, although there are exceptions to this general rule.

These beings are sometimes described as present during abduction encounters, assisting the grey aliens. I personally have no memory of such beings but plenty of abductees report seeing and interacting with them. Improbably, they may be the key to unlocking the mystery of the program.

Is This the First Time Such a Program Has Been Implemented?

Several factors argue convincingly that this is almost certainly not the first time that an enormous, planet-wide and presumably covertly executed, planned abduction/hybridization program has been instigated on a large population.

The size, longevity and complexity of the program must require a high degree of long-term strategic planning for its successful execution. The targeting of the human population entails the deployment of almost certainly a very large number of vessels specifically designed for the job of processing large numbers of abductees. These craft contain not only the equipment to perform these functions but seem to be constructed exclusively for the purpose, or else adapted and converted to the specific role. They contain grey aliens whose sole function is to execute specific tasks, and others working in the program, with instruments and tools designed for use on humans; examination tables designed and sized specifically for human bodies and other equipment deployed for ancillary procedures, such as scanning devices. As discussed in earlier chapters, the interiors of these craft are clearly designed specifically for the function of processing human abductees, creating and rearing hybrid offspring and training hybridised/hubrid adolescents and young adults. The vessels abductees are taken to are clearly not designed for any other purpose such as planetary or environmental exploration or transporting cargo of any other kind from A to B: they are obviously designed and equipped exclusively for use in the program.

One thing of note about the interiors of these vessels is that *there is a 'down'*: there is a floor, and a ceiling, and gravity precisely—or as near as makes no difference—like on Earth. None have interior spaces like the ISS, with astronauts and objects floating around in a gravity-free environment. The aliens have obviously long ago solved the issue of re-creating the gravity exerted by a normal planetary mass experienced by beings who evolved on its surface, and mastered the technology and design skills of construction for interstellar travel.

The program is clandestine, emphasising the secrecy deemed necessary by the instigators so that the target population will remain largely unaware of, or even resistant to, any idea of what is being planned and take no actions to slow down, inhibit or stop it.

There is a small error rate. UFOs are seen and filmed; abductions are occasionally witnessed; a small number of abductees retain some normal memory of the encounter; people are occasionally returned to the wrong location or wearing the wrong clothes. But the incidence of these errors is low against the huge volume of abductions, and insufficient to compromise the efficient overall operation of the program. The very small error rate is indicative of a well-practiced routine. If this were the first time such a program had been run on any population and the technology newly deployed for example, we might expect more bugs or teething troubles in the system: larger numbers of abductions may have been openly witnessed or abductees would be falling from the sky in their night clothes and killed on impact with the ground, in locations from which there is nowhere obvious from which to fall. These things don't happen, which is testimony to a well-practiced methodology with a miniscule error rate in operation. Indeed it is this minuscule error rate which, above all else, has enabled us over the years to understand most of the architecture of the program and the aliens' possible intentions: the aliens, despite their exotic futuristic capabilities, are not infallible, and *Murphy's Law* seems to apply to them as much as it does to ourselves.

What Then Are the 'Greys'?

What then of the grey aliens? What are they? Are they a different species that has been drafted into service by their mantis-like employers? Or are they workers who were somehow bred or adapted to carry out these tasks?

Talking over this issue with Dave Jacobs some years ago, he was struck by the fact that the greys have a *mouth*. Why would they need one? They don't appear to have lungs or respiratory apparatus and don't breathe air; they possess no vocal cords and never speak; and they have no digestive system so don't need (and seem physically unable) to eat. Eyes: yes, vitally important to them to see what they are doing, to lock into the optic nerves of abductees to explore memories and manipulate brain functions. Ears: yes, hearing sound is useful as deafness might be an inhibitor to personal safety and to efficiently carry out tasks. But a mouth: why would they need that? The tiny vestigial mouth is just a slit and is never seen to open under any circumstances. The mouths are never even seen to move, and the greys never display any kind of facial expression to betray emotion or mood so universally seen in humans.

For my part, I have always been puzzled by the lack of any reproductive organs. If they cannot reproduce themselves—and there is no indication that they can—then where do they come from? How are they 'born'?

The answer of course is obvious when you give the matter some thought. The mantis-aliens are masters of genetic manipulation and advanced biotechnology, as testified by the creation and gestation of hybrid alien-human offspring. The grey aliens are *manufactured*: genetically engineered specialist workers, sentient living beings designed by the mantis-beings to work at certain specific tasks.

Our current engineers very cleverly design robots, for industry and other purposes, to perform repetitive and precise tasks. Many of these robots function admirably, but they all require electrical power, and specialist maintenance such as the replacement of worn-out or broken parts. The greys, I suggest, are workers constructed by genetic engineering who will work, unceasingly, to perform precisely the tasks needed for the abduction program and to do nothing else. They have organic brains and possess high intelligence, and to some degree can respond to novel situations or crises. If an abductee occasionally breaks the neurological control which limits their movements they will smoothly deal with the situation. As organic beings, though, they require some rest/sleep time and some form of nutritional intake, but that has been accommodated for in the design.

It is possible, even likely, that the mantis-aliens did not bring the grey alien workers with them from their planet of origin but engineered them— almost certainly with the input of human sperm and ova into the zygote— here on or within the orbit of the Earth, specifically to work in the program. Hence the human characteristics: the upright bipedal body form, small vestigial mouth and nose—organs of no use to them but an inevitable by-product of biochemical engineering involving human genetic material. And they have no need for reproductive organs as *they are designed to never reproduce themselves.*

The program can probably manufacture as many workers as required, so that there will usually be replacements available for those which wear out, suffer serious accidents or fail due to other causes. They need no batteries, little sustenance or maintenance, and work ceaselessly at the tasks for which they were specifically designed: the perfect slaves. It's possible that the taller types in particular can survive only within the manufactured ecology of the craft in which they work *and never leave.* The smaller kind are obviously designed to travel down to escort the abductees up to and return them from the place (the UFO) where the procedures are carried out, as this is their primary function.

The observation that the greys are bio-engineered to not breathe air and not require an alimentary canal or digestive system of any kind leads to the conclusion that these traits are obviously of some utility in their slave workers, as the need for maintenance and supply provision (storage of food and water, atmospheric purification in the ecology where they are designed to operate) is minimised or eliminated altogether. We don't know the biological temperature tolerances of—in particular—the smaller greys, but they are never reported to wear warm or insulated clothing of any kind. The abduction of humans occurs in all climates, presumably including the polar regions. In mid-winter, in North Dakota and Minnesota, it can get pretty cold but the abductions carry on. Presumably, they have been engineered to be immune to the effects of a wide range of external temperatures.

It is possible these genetically engineered workers have a long biological lifespan relative to humans, but this lifespan is likely to be finite and they will eventually wear out and 'die', after which they may be simply replaced. However, the fact that human-alien hybrids who look 100% human on the outside are reported to be increasingly taking charge of the abduction program suggests several things:

1. The grey aliens are nearing the end of their designed biological lifespan and are wearing out

2. Not all are being replaced, as the program is entering its final stages and the primary function of these bio-engineered servants is no longer needed in such numbers

3. The abduction program was designed to run for a fixed period before its objectives are realised, i.e. the fourth or fifth generation abductee population is now sufficiently modified for whatever comes next

What Is The Origin of the Reptilian-Looking Beings and Why Are They Sometimes Reported to Be Present During Abduction Events?

If this is not the first time such a program has been run on a population, we can see where these beings may have come from: a previous successful program of hybridization on a different exoplanet in a different star system, perhaps centuries ago in our timescale. Perhaps they were brought here to assist with the early stages of the program, before the greys were bio-engineered using the sperm and ova taken from the first wave of abductees. If so, we might expect some evidence of their involvement in the early stages of the program, when the first abductees were selected from the population during the 1890s.

Does Farah Yurdozu's Story Offer a Clue?

Farah Yurdozu is a multilingual Turkish psychic and paranormal researcher whom I have known for many years. She lived for a long time in New Jersey, close to NYC, and is now based in Istanbul where I recently contacted her.

In her book, *Confessions of a Turkish Ufologist*, she relates a story recounted by her great-grandfather, Emin Refik, from the 1890s. Emin was a young man who taught at Istanbul University.

> As he stepped out of the house to go to the university, he saw a very old man smiling at him from the opposite sidewalk. The old man approached Emin and spoke to him. Though he was a stranger, the old man knew Emin's name, about his family, his job and even his financial problems. The stranger insisted that he was ready to help. He said that late in the night Emin would have some visitors, whom he should not fear. As a scientific man Emin didn't take this conversation seriously.
>
> That night he came home and went to bed. Before he fell asleep, the bedroom door opened. Two visitors entered. They were tall, and though they walked on two legs, my great grandfather later insisted they had reptilian characteristics. They wore no clothes and their skin was dark, oily and partly covered with hair. Though the stranger had told him not to fear the visitors, it was impossible not to, as they looked at him through the elliptical pupils of a snake's eyes. Emin tried to move, scream, do something … but it was impossible. He was under a total paralysis. He felt a very heavy pressure over his body. At the same time a strange vibration was filling the room. His bed started to shake. One of the reptilians got onto the headboard of the bed while the other stayed at the foot. The reptilian visitors were trying to communicate with him, and Emin fought against them mentally. He repeated again and again that they were not allowed to be in his home and commanded them to leave. He became completely unaware of the passage of time.

> At the end of the night, the two reptilians left him. While they were going they looked at him and said something about his family. **The words were spoken telepathically, directly into his mind: "From you to three generations …"** And they were gone.[11]

Yurdozu then describes how Emin Refik's five children were all psychics, the first in the family, and she herself is the third generation psychic-paranormal researcher in her family line. She speculates that these psychic abilities may have stemmed directly from Emin Refik's encounter with the reptilian beings and, by implication, whatever followed from that initial encounter.

Of her great-grandfather's encounter with the reptilian beings in the 1890s, she asks:

> What happened that night to Emin Refik is a big mystery. He was visited by two monstrous beings. Was this an abduction? Did he experience missing time? Any scars, cuts, bruises on his skin? We don't know …[12]

The significance of this encounter from Farah Yurdozu's family history seems to me to be:

1. The incident happened in the 1890s, the decade when it is probable the program was initiated

2. The reptilian beings visited the future abductee to inform him that he had been selected (by some criteria which remain mysterious)

3. Emin Refik revealed to his extended family members and to his children and grandchildren that the beings had paralysed him for the duration of their visitation

4. The beings communicated with him telepathically

5. The prophecy, *"From you to three generations,"* seems to strongly foreshadow the mystery that would bedevil countless families over the next century

6. It was reptilians, not greys, who visited him

Farah Yurdozu with the author and Peter Robbins, South Street Seaport Cafe with the Brooklyn Bridge in the background, 2008.

It is significant that Emin Refik reported no missing time, claiming full and complete recall from normal memory. From my research into this matter, it seems likely that the reptilians may have been deployed because at this early stage (the 1890s) the greys had not yet been genetically engineered to carry out the abduction program on the human population. The 'mindscan' abilities have been engineered into the taller greys, whereby they are able to store the experience of the abduction in the human abductee's long-term memory, leaving only a gap of missing time (we explored the likely neural mechanisms of this process in Chapter Seven). Perhaps the reptilian beings were from a previous program the mantis–aliens had implemented on another population, and were then brought along to help with the initiation of the program on the human population. There has never been a single report of one of the mantis–insectoids being present when a person is first abducted. It's only later, aboard their craft, at which they occasionally make an appearance. It seems that they depend on their genetic creations to handle the grunt work.

The appearance of the reptilian beings varies with reports. Although this might be due to incomplete and fuzzy memories, perhaps there is another explanation. The differences in morphology might be evidence of multiple hybridization projects on different species, or even on variations of the same species. Even here on Earth, when homo sapiens were geographically separated over large distances and were subject to differing diet, lifestyle, and environmental factors, quite striking variances in body size and appearance developed between regions. (Look at the Pygmy and the Maasai for example, or compare either with the Norse just a thousand years ago.) Whereas, in today's cosmopolitan world of comparatively easy travel, mankind has in many ways become a more homogenised species.

Farah Yurdozu's story from her family's past is a rare treasure for researchers. There exist precious few detailed accounts of encounters with such entities in the very early, pre-1900, days of the program. Few records survive save oral family histories, which slowly erode through the generations and mutate into legend, if they are not simply forgotten. My great-grandmother was born in January 1873, so by 1890 she was seventeen years old, the normal start of the reproductive years for a woman. Married in 1907, she had only one child, my grandmother, who told me she was always throughout her life taken by "the pixies." These sound like the small greys, but we never had any discussion of her detailed recollections of encounters, just confirmation that they occurred throughout her life during the one conversation we had on the subject when I was around eleven years old. Due to the geographical distance between us at that time, we saw each other only three times each year, and she died when I was twelve.

It's interesting that both Emin Refik and my great-grandmother were young adults at the same time. Perhaps the first generation of abductees forced into the program were all of similar age. Considering the dearth of information available, even extensive and diligent research is unlikely to turn up much, unfortunately. Given that most abductees today are unaware of what is happening to them at all, and most people anyway do not generally know much about their ancestors just three or four generations back, information about the first abductees chosen for the program will surely remain obscure.

The Metaphysical Mystery

If the two types of greys commonly described are in fact not technically 'aliens', but genetic hybrids containing material from the true aliens—the insect-like beings—and zygotes formed *in vitro* from human sperm and ova, specifically manufactured to work in the century-long program on the human population, then why might they choose to engineer them here in the local terrestrial ecology? Why not bring them ready-made from the alien planet of origin? Obviously the population of human abductees—the source of the genetic material—is here, but could there also be something else?

During the 1970s and 1980s, reports of tiny and sickly-looking human-hybrid babies being introduced to (especially) female abductees became widespread. These tiny babies were described as listless and unlikely to thrive. The abductees reported a mixture of reactions to them, ranging from revulsion to pity. Note that this predates reports of implanted foetuses, suggesting that these beings had developed entirely outside of the womb.

When the aliens began implanting foetuses into female abductees, often in extra-uterine sacs, the women exhibited all the normal characteristics of pregnancy. Hormonal changes included:

- Human chorionic gonadotropin hormone (hCG), a hormone which is only produced during pregnancy

- Human placental lactogen (hPL), also known as human chorionic somatomammotropin

- Oestrogen

- Progesterone [13]

After around ten weeks, the female abductee was again abducted, specifically to have the foetus surgically removed, to complete gestation *in vitro*. Why would they do this? Abductees and researchers alike have long suggested that the foetus is removed before it grows too large and the 'bump' starts to show, perhaps to keep the pregnancy from becoming known to both the abductee and others. But why implant it in the first place? They go to a lot of trouble to do this. What difference does it make to have the foetus gestate for (typically) ten weeks inside a human female, and precisely what happens during this period which then causes the hybridised creation to become more vigorous, vital, and 'human' in the way the aliens judge they need to become?

The foetuses then continue their gestation in tanks of what is presumed to be nutrient fluid. These tanks have been witnessed by hundreds of different abductees. Presumably, they are eventually 'born' when removed from the tank. Nobody has as yet reported seeing this process but it obviously happens at some point in the development cycle.

Why does 100% *in vitro* gestation fail to produce the vigorous beings they want and need for integration into human society, whereas part-gestation inside a human female seems to make all the difference?

This is well-grounded speculation, but might it have something to do with attracting and anchoring the human soul into their genetic creations?

The Soul Dimension

How does the soul (presumably free within our planetary ecology) attach itself to matter? Does it have any choice? These are metaphysical questions I feel ill-qualified to address in detail, but perhaps some readers may consider this issue and offer some ideas.[14]

Whatever the reason, it seems to make a difference and produce the beings they want and need to complete the program. We do not and cannot comprehend the level of understanding an ancient and very successful (in their own terms) civilisation might have about these metaphysical matters. Their understanding may be way beyond our current *scientific* knowledge, and moreover beyond our own moral ideas regarding genetically created beings. Surely they operate within a completely different moral and ethical framework. We just don't know and can only speculate on this issue.

The taller greys are always described by abductees who deal with them as either 'male' or 'female', despite the absence of any visible primary or secondary sexual characteristics. Might this 'soul' element determine the described gender difference? What we can say is that there must be a reason (or multiple reasons) why the grey aliens are engineered and brought into being *specifically here*, and moreover why human female gestation for, evidently, the late period of the first trimester was discovered to be so essential to the health and vigour of the young hybrid/hubrid beings needed to fulfil the alien objectives.

Regardless, it may be that this period in the program is now completed, as reproductive relationships between hubrids and their unwitting human partners might render this implantation process unnecessary. Male 'hubrids' can simultaneously impregnate a large number of human females and therefore propagate a large number of direct offspring, whereas female 'hubrids', which do reportedly exist in some numbers, may carry only one foetus at a time regardless of how many human partners they take—assuming that they are fertile and able to be successfully impregnated, which we don't know for sure.[15]

Maybe there is a different reason for the grey aliens' local manufacture, for example the logistics of interstellar transportation via whatever exotic method is employed ('Loop quantum gravity'? Hyperspace? Wormholes?—all bona fide concepts in contemporary astrophysics), which may or may not involve some of our most advanced, ground-breaking concepts in theoretical physics.[16] Darned if I know; I'm not even a physicist.

Are Other Aliens Visiting Us?

There is fairly strong anecdotal evidence of extraterrestrials visiting us but for purposes other than abducting us. For example, those reported as seen in a city park in the Russian city of Voronezh in 1989 [17] don't sound remotely like "our little friends," in size, appearance or behaviour. There have been several other highly credible reported incidents over the years around the world, none of which resemble the entities known to be involved in the abduction program.

Much of the UFO literature from serious researchers is nonetheless full of odd-ball accounts about encounters with their occupants. The English writer Timothy Good is well known for his extensive work seeking out and publishing obscure UK government documents released under the FOIA,[18] and similar from the CIA, DOD, NSA and other government sources in the United States. His recurring thesis is that a world-wide cover-up of the issue is in place from national governments and their intelligence agencies. His body of work over fifty years is impressive, and his 1987 bestseller, *Above Top Secret*, is a classic work of the genre grounded in factual accuracy. But Good also did a lot of field work tracking down weird tales like that of the 'Chupacabra' in Mexico and Puerto Rico. His books *Alien Liaison*, *Alien Base*, and *Unearthly Disclosure* contain many witness reports of encounters with apparently non-abducting entities. Tim Good, born in 1942, is still with us but now no longer active in this field of study.[19]

The French-born computer scientist and venture capitalist Jacques Vallée also followed up many cases involving encounters with entities from UFOs, some resulting in injuries and even death to the human participants. Many of these cases are detailed in his 1990 book, *Confrontations*. Vallée is first and foremost a field investigator, not an armchair theorist, and *Confrontations* contains many cases from South America to where Vallée and his wife travelled to meet and interview witnesses to these extraordinary and frequently alarming events.

However, the distinctive features of the abduction program which form the subject of this book are the ubiquity, longevity, consistency of evidence and the sheer quantity of identical testimony at its core. Not to mention the thousands of abductees over the years who demonstrate identical scarring, scoop-biopsy scars, and implants revealed by x-ray. No doubt other aliens visit us from time to time, and some of these are responsible for significant parts of the UFO phenomenon, as regularly witnessed by the world's air forces, navies and civilian pilots. But the abduction program remains unique in several respects:

1. The longevity of the program, which the evidence clearly suggests is following some plan, is now into its second century
2. Its enormous scale and global reach
3. Its fundamentally intrusive and invasive nature
4. The abductions carried out repeatedly over lifetimes with assembly-line efficiency
5. Its clear intergenerational aspect
6. The fact of the identical scoops, scars and implants found on and inside many abductees
7. The seemingly obsessive secrecy with which it is executed, which the designers of the program obviously see as essential to its eventual success

Nothing remotely like this is ever reported in connection with reported encounters with other extraterrestrial visitors. It is rare indeed to come across even two credible reports of interactions with identically described entities from different witnesses whose descriptions differ from the abducting entities engaged in the program. This does not mean that the witnesses to these other encounters cannot be believed; it means encounters with those entities *are extremely rare*. Encounters with the abducting entities, by contrast, are ubiquitous and near-identical, spanning decades.

What Is the Endgame?

No one knows the answer to this question because *they* don't tell us. Some long-term or even permanent presence here is obviously planned, because you don't run a program of this enormous size and scope, with this huge number of unwitting participants, planned so meticulously over such a long period, and executed with near 100% secrecy, without very serious intent as to the result you want and are working towards.

These beings are always reported to display no curiosity whatsoever about human societal institutions, governance or international relations. They never ask abductees anything like, "How do you choose your government?" or, "Describe to us how you as a society deal with this, or that." Similarly, there is no expressed interest in our popular culture; news media; internet or cellphone communications; or temporary preoccupations with localised conflict zones around the world. Perhaps the closest thing to that would be the occasional remarks about the degradation of our climate or planetary environment. A small minority of abductees have claimed to carry such 'messages for humanity' from their abductors. They nearly always follow a predictable script and are fortunately soon forgotten or ignored, as they are invariably the products of screen memories or other manipulations of the brain and memory of the abductee to make them believe certain things which are not true.

They don't even seem to express any interest in how human societies formally deal with health and disease matters which might affect the population, and express zero interest in any and all aspects of the global spread of the COVID 19 pandemic which they obviously see as having no relevance whatsoever to the continuance and implementation of their program.

They do not appear to favour abducting influential or high-functioning individuals, which presumably they might easily do if so motivated. Societal leaders and others elevated to high government positions, business or industry leaders or those who hold great material wealth, are of zero interest to them, as are the processes by which they've made their ascent to such positions in human societies. Instead, they stick to the genetic lines from the original abductees, all of which it appears were selected in the 1890s to either random criteria, or else if specific selection criteria were used, those still remain elusive. The living subjects of their intergenerational program, repeatedly abducted and possibly modified genetically throughout their lives in subtle ways, are not just their prime targets of concern: they seem to be their *only* concern.

Abductees are born, live their lives and die, and the program continues. Presumably at least the same proportion of abductees meet with serious or fatal accidents in life, or contract serious or fatal diseases with the same frequency as, the general human population. There are always millions more abductees moving through their lives in the program: it seems individual abductees are of little consequence to the abductors as their individual misfortunes will have no impact on the overall progress of the program as it moves relentlessly towards the fulfilment of its eventual objectives.[20]

Most abductions in recent years are reported to be by carried out by alien-human hybrids, who increasingly take charge and who seem to be—to some extent at least—replacing the greys in this function, or else augmenting them in the abduction work. There are widespread reports of integration into human society of these young adult hybrids or 'hubrids', specially bred and trained for the role. These reports are widely viewed with scepticism or else ignored and disregarded, attitudes which the designers of the program will obviously see as ideal for their purposes. The newcomers are reportedly designed to 'fit in' and pass unnoticed, and if they do fit in and in no way stand out as exceptional or odd, then everything is seen as going to plan. Again, these newcomers are reported to display no interest whatever in the human political world, its power structures or governance: they are concerned only with 'fitting in' and 'looking right', maintaining a low public profile in the society in which they are placed. Most do not reportedly even know the name of the country they now live in and display no interest in such information, which we might consider as essential in their circumstances. It's almost as if they sidestep societal institutions completely because the aliens intend to replace them with their own structures; the acclimatising 'hubrids' ignore such matters which will be irrelevant to them in the medium term, so they don't waste their time with such concerns.

They can do none of this without the co-operation of abductees, who are essential to the success of their program. If this integration is really happening, then only the abductees who are dealing with the newcomers will be able to verify the fact and extent of their presence. But due to the habitual neurological manipulation of abductees' memories, it is likely that most abductees will have no recall of the interactions without assistance in memory retrieval. The resources for such a service do not exist in any of our societies at present. The program is either unknown or its existence is not admitted. Even if such a resource were made available, the majority of abductees would likely be extremely reluctant to self-identify. The aliens seem to know instantly whether a human is in the program or not (they can always tell somehow), but in human society there is no reliable method of determining just who is an abductee, as even the program's reality is denied or ignored.

Epilogue

What is my personal attitude to this situation? Quite honestly, I wish they would pack-up and leave for good. Abandoning their project would leave us to sort out our own problems by ourselves. Serious as the seemingly intractable problems facing us may appear to be, the human race is, I believe, capable over time of agreeing and implementing solutions which deal with them. We might welcome open, voluntary assistance from a more advanced species which has been through a similar development cycle and emerged older and wiser, but the kind of covert manipulative interference evidenced by the abduction program is not the help that we need and should obviously be strongly resisted.

After a lifetime of abductions as the fourth in their hereditary line, a few years of peace without the threat of these intrusions would be welcome—and they can please remove their implants before they go. I am certainly not hostile to the idea of some alien race making peaceful—and open—contact with us; they may have several million years' evolutionary start on us, and if we could deal with them on anything like equal terms and learn from them, who would not welcome that?

It might be that some peripheral good does emerge from a forced, accelerated change from outside. Examples might be:

1. The gradual or sudden end of tribal conflict among humans

2. The permanent and irreversible dismantling of nuclear weapons and other WMDs everywhere on Earth: the aliens have repeatedly demonstrated their technical abilities to de-activate nuclear warheads and delivery systems at distance [1]

3. The accelerated ending of the outdated and tribalism-fostering 18th century concept of 'the nation state'

4. A managed reduction in the numbers of the human population, either drastic or gradual, to a level ecologically sustainable

5. A less toxic planetary environment free of industrial and plastic pollutants of the atmosphere, waterways, and oceans

6. An end to, and possible reversal of, catastrophic climate change caused by human societal and industrial activity powered by the CO_2-generating petrochemical industry

But it looks like that option is not going to be on offer from "our little friends." They are evidently focused entirely on their own scheme, which appears to be planetary acquisition by species-absorption through a long project of covert, involuntary hybridization using a small percentage of the local target population. It's not exactly the gift to humanity that the New Agers would have us believe is their purpose in visiting us here on Earth.

Like almost all 'aware' abductees, I have never wanted any of these intrusions into my life. It's now almost impossible to imagine how it might have proceeded if everything were 'normal', and this *secret life* were not going on all the time under the surface.

So Where Do We Go From Here?

The abject failure of the mainstream scientific/medical communities, and those throughout the human world placed in governmental positions of responsibility in human societies, to even recognise—let alone take seriously—the existence of the abduction program, its scale, longevity and possible purposes is frustrating to say the least. Despite the extravagant claims of conspiracy theorists, the available evidence demonstrates that not much is genuinely known or understood about the matter. We don't know what we don't know, but it seems apparent that even those in leadership positions, with access to information about the ET/UFO issue at the highest levels, still know very little about it. Despite the recent revelations of the Pentagon's AATIP program in December 2017, we may be reasonably confident only that:

1. The UFO/UAP phenomenon is acknowledged[2]

2. Much speculation still abounds about whether these craft were built on Earth! Yesterday it was the Soviets; now it's China. Many highly classified films of UFOs/UAPs in flight have been collected over many decades but they tell us nothing whatsoever about the intentions of their crews

3. Credible Evidence that the abduction program is understood or even acknowledged by any official sources remains rather thin. And it appears that the program, after more than a century with no serious human interference, may now be in its final stages

Are We Now Out of Time?

Unfortunately, I personally think the answer to this question is affirmative.

The aliens' regular activities in carrying out the abduction program display some remarkable technology. They routinely make both themselves and their craft invisible to the human eye and much, or all, of our sensor apparatus. They can somehow temporarily affect or suspend the atomic structures of both objects and people, enabling both the abductors and abductees to pass unharmed through solid walls, floors, and closed windows. People are carried into the UFOs on 'beams of light' and remain invisible to onlookers. And getting here in the first place, in enormous numbers—while successfully concealing their presence from the human population at large—is a stupendous achievement against our current notions of what might be technologically possible. These technologies are way beyond our current capabilities to replicate or match; for them, by contrast, they look to be routine.

Abductees are made to forget all details of the experience, retaining only an awareness of missing time and a deep sense of unease or puzzlement. This is advanced physics and neurology at work: they are very experienced at this. The processes are routinised and the whole gigantic operation is well-practiced, planned and executed with a thoroughness which leaves very little to chance. It is not possible to know the natural lifespan of the aliens but they seem to have a long timescale, moving relentlessly and thoroughly towards the conclusion of their project, the timespan of which obviously spans several human generations.

The neurological powers of the aliens seem to be supernaturally beyond what we have ever imagined possible in any species. They live in a completely telepathic society and can individually control the perceptions and behaviour of any less-developed species at distance. Close up, they can implant thoughts, induce amnesia, and reconstitute memories. These abilities are inherent at a biological level, not dependent on some external technology deployed and utilised at arm's length: this makes a difference. Moreover, the evidence seems to demonstrate them to be masters of bio-genetic engineering. Given time, they are *able to change and bio-engineer other species in whatever ways they choose*. It looks as though they habitually choose for other species to become permanently subservient to them. Moreover, they take as much time as they need and are very thorough. We might not like what they choose for us, but it looks like we ourselves may be offered no choice in the matter.

During the past forty years there have been ample opportunities to recognise the program, alert the global population, and take some at least partly effective measures to slow down or even stop the progress of the abductors' plans. The undeniable evidence that total secrecy about the program has been maintained by the aliens strongly suggests the potential vulnerability of their plans should their intentions be discovered. Unfortunately, the scientists, technologists, and medical professionals working in the human world have chosen to remain ignorant, 'sceptical' of the experiences of those caught up in the abduction program and deaf to what is reported by the comparative few who have become aware of what is happening to them. In what are possibly its final stages, the program remains unacknowledged or little understood. Decades of credible, and nearly identical, testimonies of thousands of abductees continue to be disbelieved. The aliens may achieve their intended results with the generous but unwitting assistance of the larger part of the human population that seems determined to ignore the danger they are facing.

If the abduction program is close to achieving its purpose, perhaps all we can do is wait and see what the future holds in store for us. Whether the present generation of abductees—possibly 2–5% of the global population—will share that future with the rest of humanity is unknown. Are we destined to go our separate ways, as it were? What's in store for those who've been made a part of this program? And what of all the others?

There have been numerous dark forecasts throughout history of an apocalyptic ending for the human race. The Abrahamic religions are full of such dire prophecies, whether the Second Coming of Christ, the rise of the so-called 'Antichrist', or the arrival of the Twelfth Imam of the Shi'ite sect of

Islam. Apocalyptic warnings abound of the catastrophic consequences of climate change; of the threat of annihilation from global thermonuclear warfare; of financial meltdowns and societal collapse. No one wants to hear *yet another* dire apocalyptic prediction: that a race of extraterrestrial visitors has been executing a covert program of subtle genetic modification of a small percentage of the human race for more than a century with the prime objective of quietly taking over control of human societies on Planet Earth. From the paradigms of our current consensus reality, this one seems preposterous: *I just **cannot** believe it.*

I have every sympathy with this position. With all our present preoccupations and concerns, most people understandably pay no attention to even the possibility. And those who do feel helpless as individuals in the face of governmental, intelligence, and scientific resources that remain stubbornly indifferent to the matter. It should be their prime responsibility, after all, but if they are giving it no attention ... well then, it can't be real, can it?

As explained earlier, if these beings would just choose to abandon their program, pack up and leave, then no one would be happier than I. If no one in future generations even believes that this program ever existed or was prosecuted throughout the 20th century, that would be fine by me so long as they were *gone*.

However, all the signs are that the aliens are not likely to abort their program at this stage. A lot of planning and resources have been invested in this enterprise: it may now be too late to effect any material change to the outcome.

We may be *out of time*.

Appendix A

Letter from Dr. Joel Kassimir, Manhattan Dermatopathologist, concerning the scoop mark scar biopsied and analysed on 11 December 2008 [1]

ACKERMAN ACADEMY OF DERMATOPATHOLOGY
145 EAST 32ND STREET • 10TH FLOOR • NEW YORK, NEW YORK 10016 • TEL 212.889.6225/800.553.6621 FAX 212.889.8268
DERMPATH DIAGNOSTICS

REPORT

accession no. **AS08-120664**
patient **Aspin, Steve**
dob/age April 16, 1956 (52) sex Male

Joel Kassimir, M.D.
10 East 88th Street
New York, NY 10128

date of biopsy December 11, 2008
date specimen received December 12, 2008
date reported December 15, 2008

tel. 212-876-3319
fax 1-212-423-0840

ANATOMIC SITE
Right posterior shoulder

CLINICAL INFORMATION
"Dermatofibroma"

[handwritten: Steve — I hope by now you've already received this by FAX —Bob]

DESCRIPTION OF FINDINGS BY EXAMINATION GROSSLY
The specimen is received in formalin and consists of a punch biopsy of skin measuring 0.4 x 0.4 x 0.3 cm. The specimen is bisected and submitted in toto in one cassette to be sectioned.

DESCRIPTION OF FINDINGS BY EXAMINATION HISTOPATHOLOGICALLY
An increased number of fibrocytes and histiocytes is accompanied by coarse bundles of collagen arranged randomly.

DIAGNOSIS

RIGHT POSTERIOR SHOULDER - DERMATOFIBROMA

Yushan Chiou
Yushan Chiou, M.D.

YC/jhb

A. Bernard Ackerman, MD • Erika M. Balfour, MD • Yushan Chiou, MD • Geoffrey J. Gottlieb, MD
Ying Guo, MD • Patricia A. Heller, MD • Sandra Mass-Cohen, MD • Howard L. Martin, MD • Joan M. Monti, DO

[handwritten: results delayed]

New York PFI# 7444 CLIA# 33D0961530

Appendix B

Images of Bodily Scarring, Puncture Marks and Bruising following Abductions

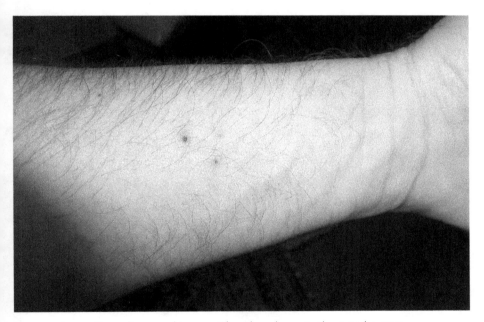

Arm punctures - these discovered in 2010. There have been similar over the years.

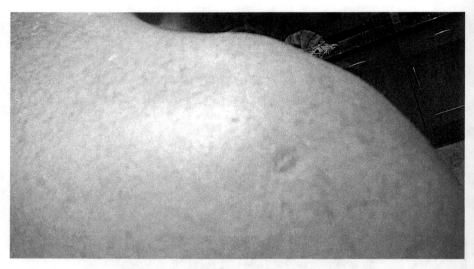

Scoop mark on right shoulder - from August 2008. Biopsied by Dr. Kassimir in December 2008, as detailed in Chapter Four.

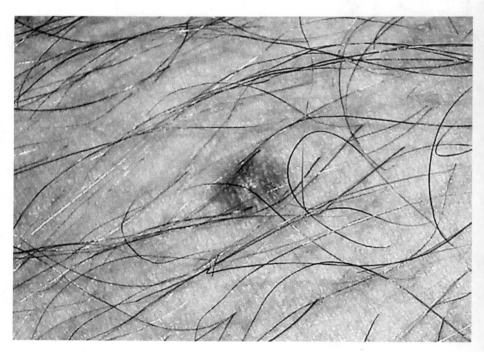

Scoop mark behind right knee, of indeterminate vintage.

Appendix B | 237

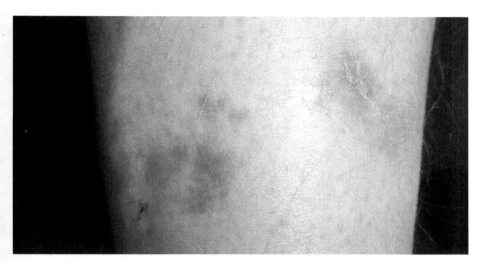

Leg burn from July 2013.

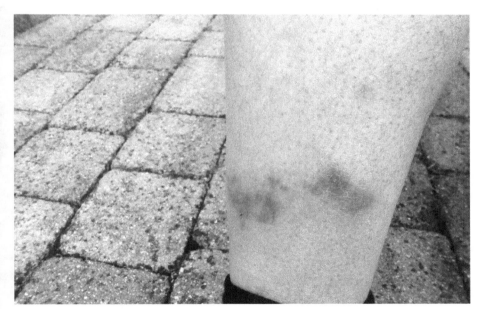

Leg burn from 2013.

Appendix c

Forensic Laboratory Analysis by Microtrace LLC of Hairs found following Two Abduction Events in 2015

Below is the September 2016 report from Microtrace LLC,[1] conducted by Senior Research Microscopists Jason C. Beckert and Skip Palenik.

The report concerns the biochemical analysis of the two hair samples:

- From a female abductee in California (unknown to me) following her reported interaction with a female hybrid being during an abduction event in 2016

- Recovered from the rug beneath our dining room table, the day following the abduction event described in Chapter Four in the section sub-headed 'Hair of the Alien'

The results demonstrate that these two hairs, retrieved from abductees living thousands of miles apart and unknown to each other, were both probably from manufactured wigs.

The first sample analysed, from CA, was found to be a pigmented polypropylene fiber commonly used in inexpensive costume wigs.

The second was identified as a human head hair, but bleached and with traces of blue pigment, as with a cosmetically treated wig made from human hair. This conclusion is consistent with my memories of the appearance of the abducting entity which entered the house through the garden doors, a drawing of which appears in Chapter Four.

Microtrace LLC   microscopy • microchemistry • forensic consulting

07 September 2016

RE: MT16-0183 – Analysis of two potential hairs

We have completed our analysis of the two items submitted to our laboratory for identification. The analytical methods employed, the results obtained, and the conclusions we have drawn from them are detailed in the report that follows.

Samples

The following samples were received at our laboratory on 26 July 2016 (Figure 1):

- Q1 – "Brown Hair, 2006 Found on bare shoulder beneath BXXXXX's night shirt" (Figure 2)
- Q2 – "White Hair, 2014 Found on floor of home SXXX and SXXX" (Figure 3)

Task

- Identify the items.

Analytical Approach

Q1

A single black fibrous item (~ 5.7 cm long) was located on the adhesive of a folded Post-It® note inside the Q1 packaging (Figure 4). It was transferred to a microscope slide and examined using polarized light microscopy (PLM). The item is a synthetic fiber (~ 95 – 100 μm in diameter) with moderate birefringence and a positive sign of elongation (*i.e.*, it is not a hair) (Figure 5). Its color is primarily the result of fine (< 1 μm) black pigment particles, although trace amounts of blue pigment were observed as well (Figure 6).[1] The fiber is slightly twisted along its long axis which has likely imparted some irregularities to its approximately circular cross-sectional shape (Figure 7). It exhibits no visible fluorescence using ultra-violet (UV) and blue excitation and only a very faint red fluorescence when viewed with green excitation.

---

[1] Due to the relatively high density of pigment, individual pigment particles are best observed when a portion of the fiber is pressed into a thin film.

790 Fletcher Drive, Suite 106 • Elgin, IL 60123-4755 • 847.742.9909 • Fax: 847.742.2160
www.microtracellc.com

A short segment of the fiber was analyzed using a microscope equipped with a hot stage in order to determine its melting point. The fiber becomes isotropic (*i.e.*, it reaches its crystalline melting point) at ~ 166°C.

A portion of the fiber was mounted on a polished salt plate and chemically analyzed using Fourier transform infrared microspectroscopy (micro-FTIR). It was identified as polypropylene (PP) (Figures 8 and 9). This is consistent with the optical and physical properties previously described.

The fiber was also analyzed using Raman microspectroscopy (micro-Raman). Carbon black and polypropylene were identified throughout the fiber (Figure 10). The blue pigment was identified as a phthalocyanine pigment (Pigment Blue 15 – beta polymorph) using confocal sampling (Figure 11).

A portion of the fiber was washed in hexane, carbon coated, and elementally analyzed using energy dispersive x-ray spectroscopy (EDS) in a scanning electron microscope (SEM). Carbon (C) was the only element detected (Figure 12). This is not surprising given that is it almost entirely composed of polypropylene and carbon black.

Q2

A single white fibrous item (~ 6.0 cm long) was located on the adhesive of a folded Post-It® note inside the Q2 packaging (Figure 13). It was transferred to a microscope slide and examined using PLM. The item is a colorless hair (~ 90 – 100 μm in diameter) although it has faint blue coloration in some areas (Figures 14 and 15). This coloration is due to the presence of discrete blue pigment particles on the surface of the hair (Figure 16). The hair does not have a root and it has an irregular cross-sectional shape (Figure 17).

The cortical texture of the hair suggests that it has been cosmetically treated. This finding is supported by the strong staining when the hair is placed in a solution of methylene blue and by the pronounced fluorescence when it is examined using UV, blue, and green excitations (Figures 18 and 19).

A portion of the hair was washed in hexane, carbon coated, and then elementally analyzed using SEM-EDS. It is composed of carbon, oxygen (O), nitrogen (N), sulfur (S), and calcium (Ca) (Figure 20). SEM also shows that the cuticle has been damaged (Figure 21).

The blue portions of the hair were examined using micro-Raman but no additional information was obtained.

Summary and Conclusions

Two fibrous items were submitted to our laboratory for identification. One of these samples (Q1) is a pigmented polypropylene fiber. Its intended use cannot be stated with certainty but its large diameter (~ 95 – 100 μm) suggests that it may have originated from a wig. Typically, only inexpensive (*e.g.*, costume) wigs use polypropylene fibers.

MT16-0183

Q2 is consistent with a human head hair. It is colorless although it appears to have been cosmetically treated (*i.e.*, bleached). There is a small amount of blue pigment on some areas of the hair. Aside from their similar diameters, there is no obvious connection between Q1 and Q2.

If you have any questions concerning this report, or if we may be of further assistance, please do not hesitate to contact either of us directly. Thank you for consulting Microtrace.

Sincerely,

Jason C. Beckert
Senior Research Microscopist

Skip Palenik
Senior Research Microscopist

*This report shall not be reproduced except in full, without written approval of Microtrace.
Analyses performed at Microtrace are accredited under ISO/IEC 17025.
See certificate #AT-1932 issued by the ANSI-ASQ National Accreditation Board.*

Microtrace ʟʟᴄ

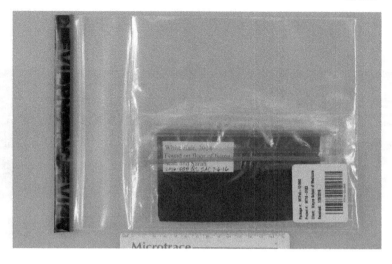

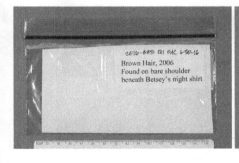 

Figure 2. Q1 sample as received.

Figure 3. Q2 sample as received.

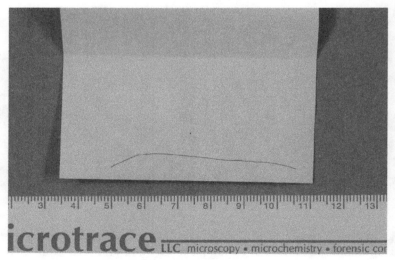

Figure 4. Contents of sample Q1.

Figure 5. Q1 fiber mounted in xylene and viewed using transmitted light (left) and between crossed polars with a first order red compensator inserted (right).

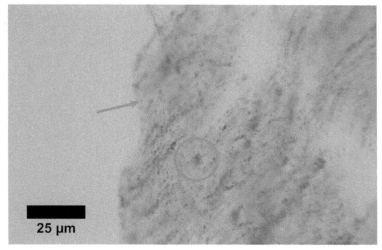

Figure 6. Discrete black and blue pigment particles in a pressed portion of the Q1 fiber (mounted in xylene and viewed using transmitted light). A single blue pigment particle is marked with an arrow while an undispersed agglomeration of blue pigment is circled.

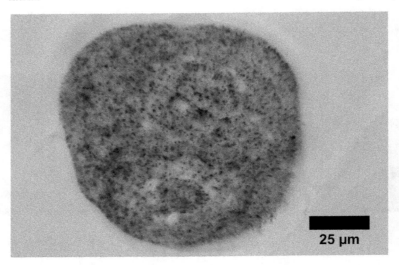

Figure 7. Cross-section of the Q1 fiber (mounted in xylene and viewed using transmitted light).

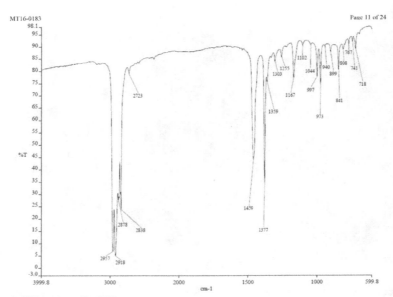

Figure 8. FTIR spectrum of the Q1 fiber.

APPENDIX C | 247

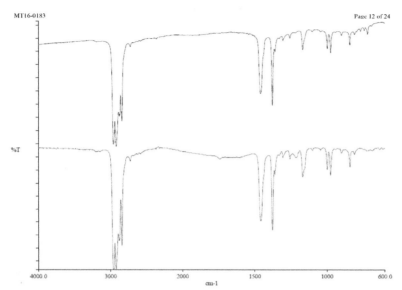

Figure 9. FTIR spectrum of the Q1 fiber (top) compared to a reference spectrum of polypropylene (bottom).

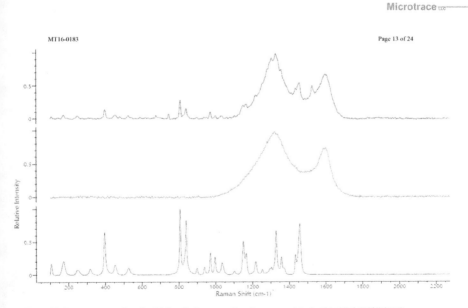

Figure 10. Raman spectrum from the Q1 fiber (top) compared to reference spectra of carbon black (middle) and polypropylene (bottom).

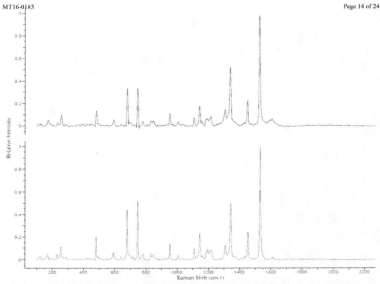

Figure 11. Raman spectrum of the blue pigment from the Q1 fiber (top) compared to reference spectrum Pigment Blue 15 (beta polymorph) (bottom).

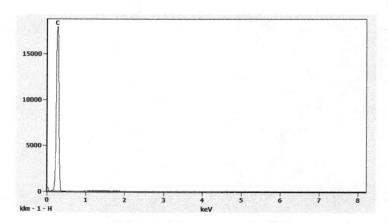

Figure 12. EDS spectrum of the Q1 fiber.

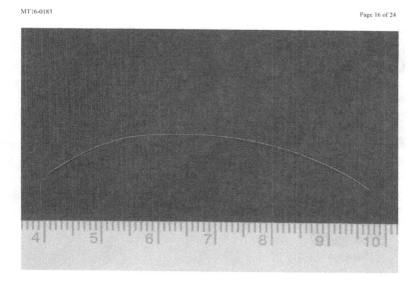

Figure 13. Q2 sample after it was removed from the Post-It® note.

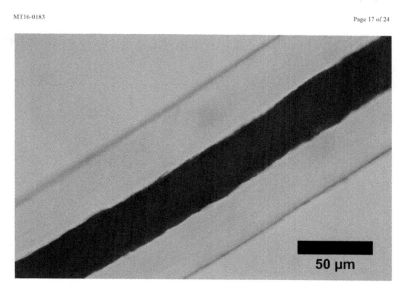

Figure 14. Typical portion of the Q2 hair (mounted in xylene and viewed using transmitted light).

Figure 15. Portion of the Q2 hair with a slight blueish coloration (mounted in xylene and viewed using transmitted light).

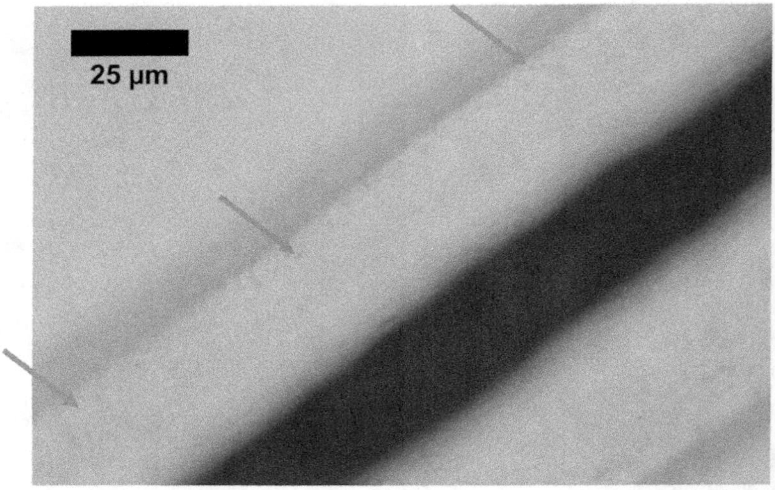

Figure 16. Discrete pigment particles (a few of which are marked with arrows) on the surface of the Q2 hair in an area with a slight blueish coloration (mounted in xylene and viewed using transmitted light).

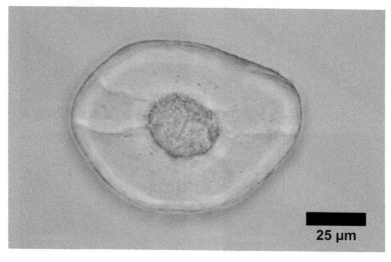

Figure 17. Cross-sectional view of the Q2 hair (mounted in xylene and viewed using transmitted light).

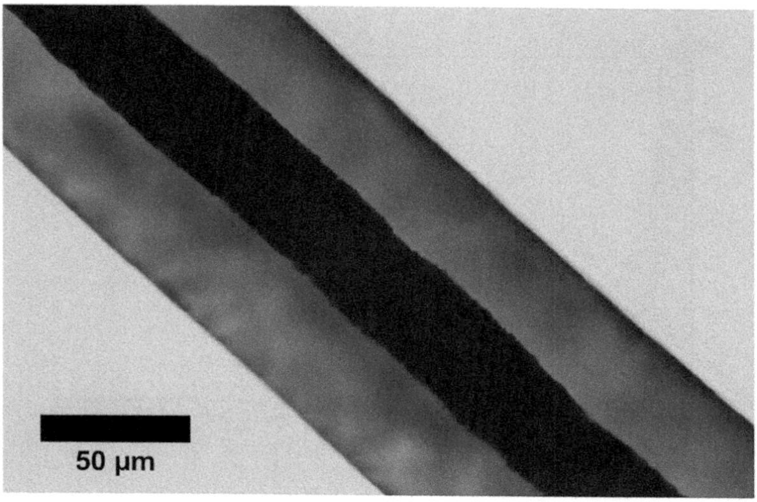

Figure 18. Portion of the Q2 hair after staining with methylene blue (mounted in xylene and viewed using transmitted light).

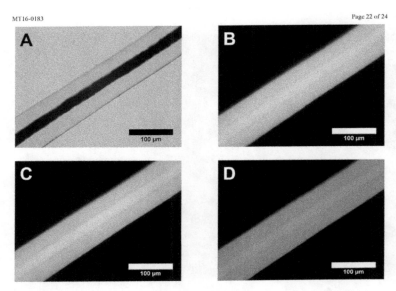

Figure 19. Q2 hair viewed using A) transmitted light B) UV light C) blue light and D) green light (mounted in xylene).

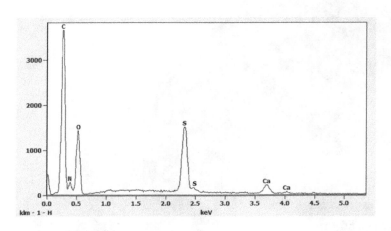

Figure 20. EDS spectrum of the Q2 hair.

Appendix C | 253

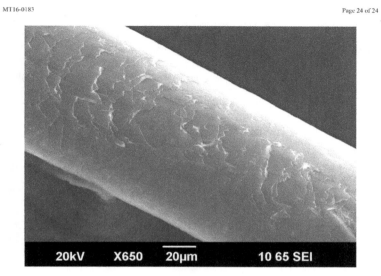

Figure 21. SEM image of the Q2 hair showing the damaged cuticle.

Appendix D

Materials Analysis of Exotic Nano-Device Implant, Surgically Removed from Patient 'John Smith' & published by Steve Colbern on 25 January 2009

Below is a distillation of the report by materials analyst Steve Colbern of an implanted nano-device surgically removed from patient 'John Smith' (a pseudonym) by Dr. Roger Leir and Dr. John Matriciano on 6 September 2008. The first part describes when and how 'Mr. Smith' discovered the presence of the device, confirmation of its presence by x-ray and CT and its subsequent removal by Leir's surgical team.

There follows a detailed analysis of the composition and properties of this nano-device as reported by Colbern on 25 January 2009. This section is probably only of serious interest to those technically qualified and practising in the materials sciences, as many of the terms and details are of a highly technical nature.

The report's findings are summarised in Chapter Four, which contains the summary of a dialogue between Colbern and Robert Hastings about this object. Hastings himself has qualified expertise in the analysis of semiconductors, so the dialogue is highly pertinent and helpful in demystifying aspects of the report which may otherwise remain obscure to the reader unqualified in these highly complex areas of materials science.

The takeaway is: the device is almost certainly a complex manufactured nano-device implanted in this patient, of probable extraterrestrial origin but of unknown function or purpose.

The full report may be viewed unedited at:

> https://docplayer.net/76207851-Analysis-of-object-taken-from-patient-john-smith-report-author-steve-colbern-25-january-2009.html

Analysis of Object Taken from Patient John Smith

Report Author: Steve Colbern
25 January, 2009
© *by S. G. Colbern. All Rights Reserved.*

Background Information

Personal History of Subject

Mr. John Smith is in his 40s, and is married with 3 children. The subject has a strong technical background, and is currently a researcher in the Materials Science field.

Case History

The subject has had a lifelong history of UFO sightings and missing time experiences. On the night of February 28, 2008, Mr. Smith was alone in his house. About 10:30 PM, he sighted two unusually large raccoons in the avocado tree, in his back yard. He fed the animals, and observed for them for some time. Mr. Smith then went to bed, and slept until approximately 8:00 AM the following morning.

Upon awakening, Mr. Smith noticed a burning pain in the tip of his left, second toe, and a soreness on the right side of his head. Inspection revealed two apparent puncture wounds on the underside of the end of the left, second, toe, and a scratch on its right side. One of the puncture wounds in the toe was found to fluoresce green under ultraviolet (UV) illumination.

Over the next four days, the pain in the toe increased, and felt like a strong electric shock, whenever any weight was placed on the end of the affected toe. The pain was at a maximum four days after the incident, and decreased slowly thereafter.

Mr. Smith then saw Dr. Roger Leir, who obtained x-rays of Mr. Smith's left foot. A small (~3 mm) foreign object showed up on the x-rays, under the end of the distal phalanx bone of the left, second, toe (Figure 1).

The object resembled a bent piece of wire on the x-ray. Dr. Leir commented that the object had approximately the same x-ray density as human bone.

A subsequent CAT scan of the left foot confirmed the presence of a foreign object in the same toe.

Gaussmeter, and radio frequency analyzer (RF) tests were done on the object, on August 21, 2008, by Dr. Leir, at his Thousand Oaks office, while it was still in Mr. Smith s body. These tests indicated that the object was emitting radio waves in the Gigahertz (1.2GHz), Megahertz (110MHz and 17 MHz), and Extremely Low Frequency (ELF, 8Hz) bands. The object also generated a magnetic field of > 10 mgauss.

The object in Mr. Smith s toe was removed surgically on September 6, 2008, by Dr. Roger Leir, and Dr. John Matriciano. The object was apparently brittle, and broke into 12 pieces during removal. The shape of the object was originally cylindrical, with a size (~4 mm × 1 mm dia.) and shape very similar to several objects that Dr. Leir had removed previously, from other patients (Figure 2).

Pathology tests on the tissue surrounding the object showed no inflammation or immunological reaction by the subject's body to the presence of the object.

The pieces of the object turned black, then red, upon refrigerated storage in blood serum, taken from Mr. Smith. Within 12 hrs of removal, the pieces of the object lined up in the original order, as if trying to re-assemble.

One of the pieces of the object was given to the author of this report for analysis. Mr. Smith stated that after the object was removed, there was a definite, but subtle change in his mood and thought processes, and that he felt more like his 'old self'.

Analytical Procedure

The sample was inspected with the naked eye, and then imaged under light microscopy, using an Olympus SZ-40 dissecting scope and an Olympus BH-2 microscope. Magnifications from 10×–400× were utilized.

The sample was then imaged using a JEOL 7500F scanning electron microscope (SEM). Magnifications between 150× and >100,000× were utilized. The sample

258 | OUT OF TIME

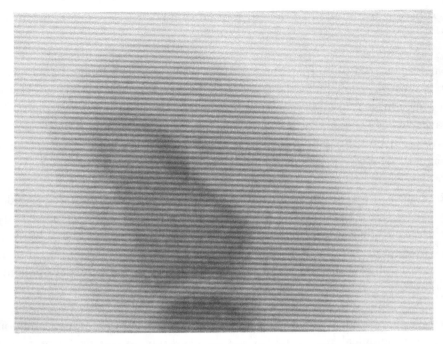

Figure 1: X-ray Image of Mr. Smith's Left, 2nd Toe Showing Foreign Object Figure

Figure 2: Implant Object Removed Intact, similar to John Smith Object before removal

was subjected to energy dispersive x-ray elemental analysis (EDX), along with the SEM imaging.

The sample was then exposed to a strong magnetic field, generated by a neodymium-iron-boron (NIB) magnet, to observe any ferromagnetic behavior.

Raman spectroscopy was also done on the sample, using a Horiba/Jovin Yvon Aramis Raman spectrometer. Both 532 nm and 633 nm laser excitation wavelengths were used.

The sample was later (26 November, 2008) sent out to an independent laboratory for more sensitive elemental analysis, using inductively coupled plasma-mass spectrometry (ICP-MS).

Analysis Results

Appearance of Sample
The sample was a small chunk of solid material, approximately cubic in shape, approximately 1 mm × 0.5 mm × 0.5 mm in dimensions, and dark, reddish in color. The sample was delivered stored in a small, plastic, screw-top medical specimen vial, and covered in blood serum from the patient, to prevent degradation.

Light Microscopy
The sample was first imaged under the dissecting microscope, to obtain lowmagnification views (10× – 40× magnification) of the entire piece. The sample was imaged both in the original container, in blood serum, and in the open air (Figures 3 and 4). Drying the sample appeared to cause no degradation. The low-magnification views of the sample revealed that the sample had a somewhat rough, and irregular, surface.

A reddish patina extended over a large percentage of the sample surface, which had a color resembling that of iron oxide (Figures 5, 6, and 8). The patina also strongly resembled the corrosion product seen on the surfaces of iron meteorites, which have been exposed to the Earth's atmosphere for some time (Figures 9 and 10).

Figures 3 and 4: Sample in Blood Serum and in Air – 10× Magnification

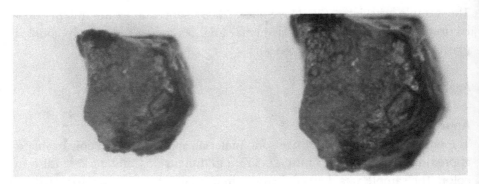

Figures 5 and 6: Sample in Air – 30× and 40× Magnification

Figures 7 and 8: Sample in Air – 40× Magnification

A dark, shiny, surface was also observed on the surface of one side of the sample, which remained shiny even after the sample had completely dried (Figure 6). The sample was then placed under the high-power Olympus microscope, and images taken at 50× – 400× magnification (Figures 11–16).

APPENDIX D | 261

Figures 9 and 10: Fragment of Campo Del Cielo Iron Meteorite 10× and 20×

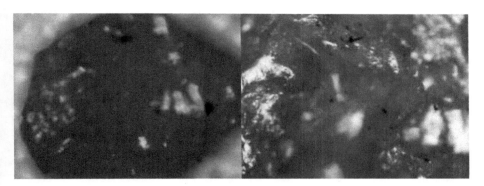

Figures 11 and 12: 50× and 100× Magnification of Sample – Showing shiny surface and inclusions of white material

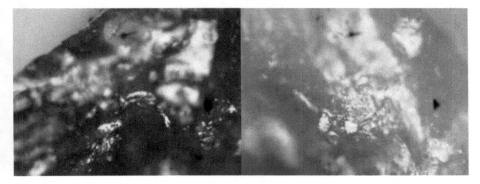

Figures 13 and 14: 100× and 200× Magnification of Sample

Under higher (50× – 100×) magnification, inclusions of a light-colored material were evident on the dark, shiny side of the sample (Figures 11 and 12). At 400× magnification, an opalescent sheen was revealed on the shiny material,

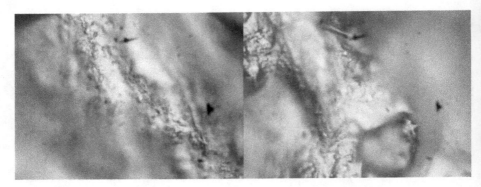

Figures 15 and 16: 400× Magnification of Sample – Showing opalescence of shiny surface material

which resembled that of mother-of-pearl (Figures 15 and 16). Areas of red patina were interspersed with the areas of shiny, opalescent material.

SEM Imaging

Scanning Electron Microscope (SEM) imaging was done with gradually increasing magnification, with the first images being taken at a magnification low enough to show any bulk structure possessed by the sample (Figures 17 and 18).

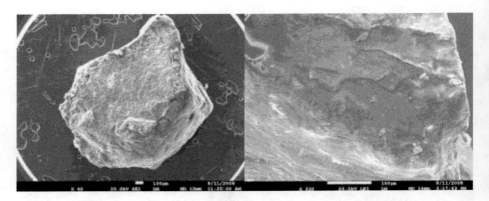

Figures 17[1] and 18: Low magnification Images of Sample (40× and 230×) – Showing whole sample and outer layer over darker bulk material

---

1   Figure 15 shows the shiny sample surface, seen in light microscopy (upper sample surface in image in Figure 5).

APPENDIX D | 263

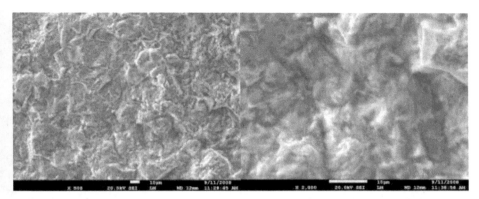

Figures 19 and 20: Higher Magnification Views of Shiny Layer of Sample (500x and 2,000x)

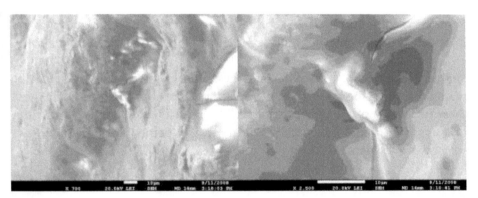

Figures 21 and 22: Higher Magnification Views of Dark, Bulk, Portion of Sample (700x and 2,500x)

Figures 23 and 24: Inclusions of Light Material in Dark, Bulk, Area of Sample (1,100x and 2,700x)

These images showed that the shiny, opalescent, phase seen in the light microscope images appeared to represent an outer layer, or coating on the sample (Figure 18). The inner, bulk, portion of the sample consists of a darker[2] material.

The material that the outer layer of the sample was composed of was seen to be fairly rough, under 500× – 2,000× magnification, with the largest surface irregularities on the order of a 5–20 microns (µm) in height (Figures 19 and 20).

The darker, inner, bulk, material of the sample appeared somewhat smoother than the sample outer layer, with surface irregularities on the order of a few microns in height (Figure 21).

Some areas of the bulk material appeared quite smooth, with few surface features in evidence, other than cracks, and some light-colored areas (Figure 22).

Higher magnifications revealed that these light-colored areas resemble the material seen in the outer layer of the sample (Figures 23 and 24). Some of these inclusions of the lighter material had unusual, and complex, structures (Figures 25, 26, and 33), with sharp, horn-like, points, and long bone-like structures in evidence.

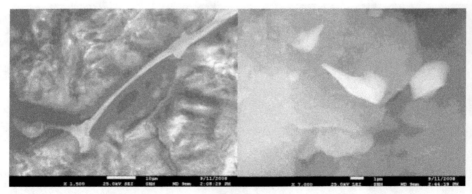

Figures 25 and 26: Views of Oddly Shaped Light Material Inclusions (1,500× and 7,000×)

---

2 Dark, in the context of SEM imaging refers to a surface which absorbs electrons efficiently. Light areas in SEM images are those which reflect electrons efficiently. In this case, the darker material, seen in SEM images, was the Fe–Ni phase (see EDX data).

APPENDIX D | 265

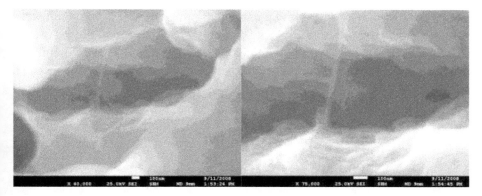

Figures 27 and 28: High Magnification Views of Cracks in Shiny Layer of Sample – Showing Nanofibers (40,000× and 75,000×)

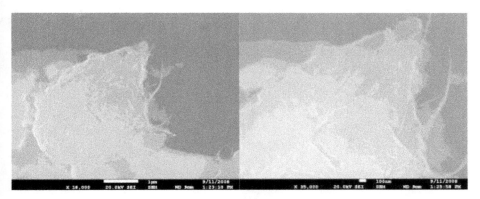

Figures 29 and 30: Higher Magnification Views of Light Material Inclusion in Dark Area of Sample – Showing Nanofibers (18,000× and 35,000×)

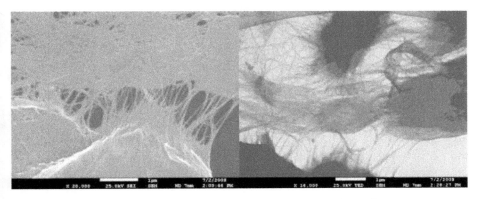

Figures 31 and 32: Single-Walled carbon Nanotubes (arc process) – shown for Comparison

Nanofibers, with a primary bundle diameter of approximately 10 nanometers (nm) were seen in both the sample outer coating material, and in the inclusions of light material in the dark areas of the sample (Figures 27–30). These nanofibers resemble bundles of single-walled carbon nanotubes (SWCNT, Figures 31 and 32).

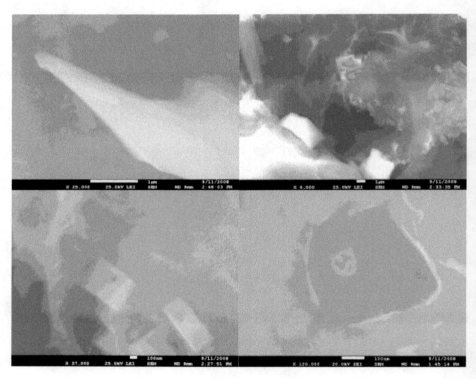

Figures 33–36: Unusual Surface Structures in Bulk Material of Sample

Highly regular crystal inclusions, and small pits, both ~500 nm in largest dimension, were also seen in the dark areas of the sample (Figures 33–36).

EDX Data

Energy Dispersive x-ray (EDX) elemental analysis scans of a relatively large area of the sample[3] (500× magnification) revealed that the object is composed mainly of oxygen, iron, nickel, carbon, and silicon.

---

3 The sample was labeled Sun Nano Impurity, in the EDX data, for security reasons.

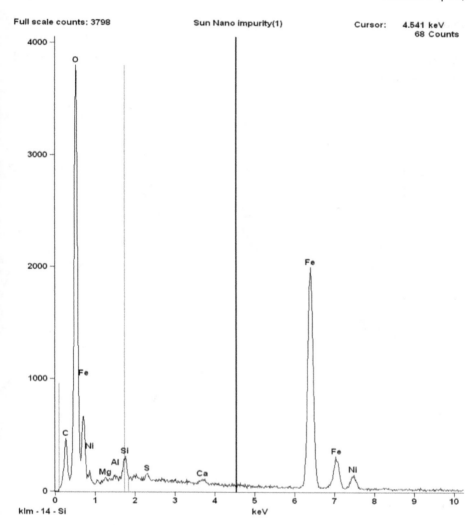

Figure 37: Large-Area EDX Scan of Object

Smaller amounts of magnesium, aluminum, calcium, and sulfur, phosphorus, and sodium also showed up in the initial EDX scans (Figures 37 and 38).

The small-area EDX scan (point and shoot technique), in which the elemental composition of a very small area in an SEM image is analyzed, was used to determine that the darker areas of the sample are composed mainly of iron and nickel, with a high content of carbon and oxygen.

Small amounts of silicon, along with traces of sodium, magnesium, aluminum, phosphorus, sulfur, chlorine, calcium, were also present in the dark material (Figures 39, 40, 43, 44, and 48). The EDX instrument software also detected traces of tungsten, and iridium (not shown).

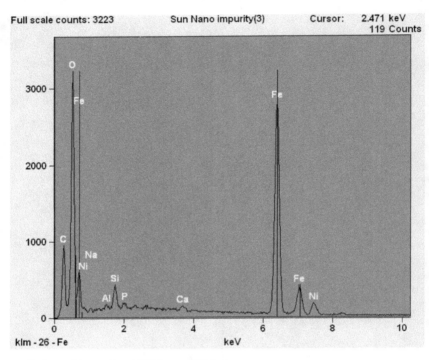

Figure 38: Second Large-Area EDX Scan of Object

The Fe/Ni peak height ratio data was used to calculate the percentage of nickel[4] in the Fe–Ni phase of the sample. The concentration of nickel was found to be 5 wt. % – 6 wt. % with respect to the mass of the Fe–Ni sample phase.

The lighter areas of the sample are composed mainly of carbon, oxygen, silicon, sulfur, aluminum, calcium, iron and nickel, with smaller amounts of sodium, phosphorus, chlorine, potassium, and titanium (Figures 39–42, and 44–45).

Much lower amounts of iron and nickel were detected in the light areas of the sample than in the dark areas.

---

4 Calculated as Ni peak height / (Fe peak height + Ni peak height)

APPENDIX D | 269

Sun Nano impurity(2)

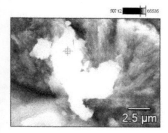

Image Name:
Sun Nano impurity(2)
Accelerating Voltage: 20.0 kV
Magnification: 7000

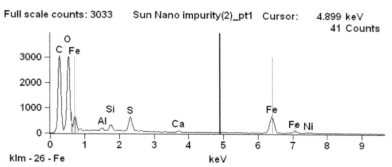

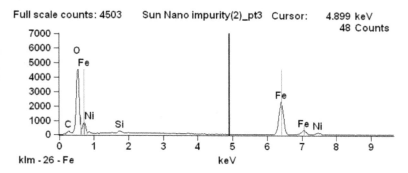

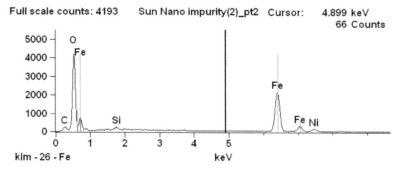

Figure 39: Small Area EDX Scan of Light and Dark Areas of the Sample

Sun Nano impurity(4)

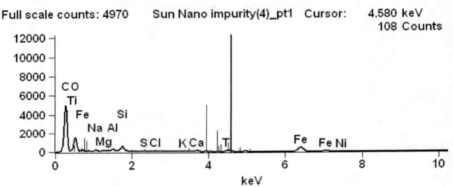

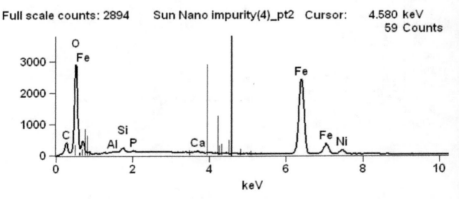

Figure 40: Small Area EDX Scan of Light and Dark Areas of the Sample

Sun Nano impurity(5)

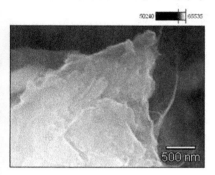

Image Name: Sun Nano impurity(5)

Accelerating Voltage: 20.0 kV

Magnification: 35000

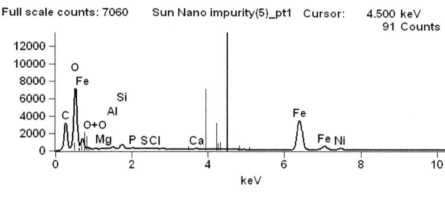

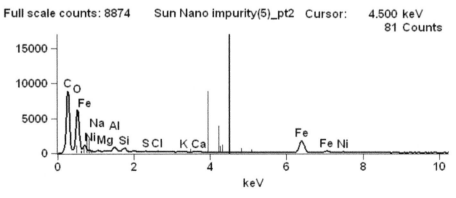

Figure 41: Small Area EDX Scan of Light Area of the Sample Showing Nanotube Bundles

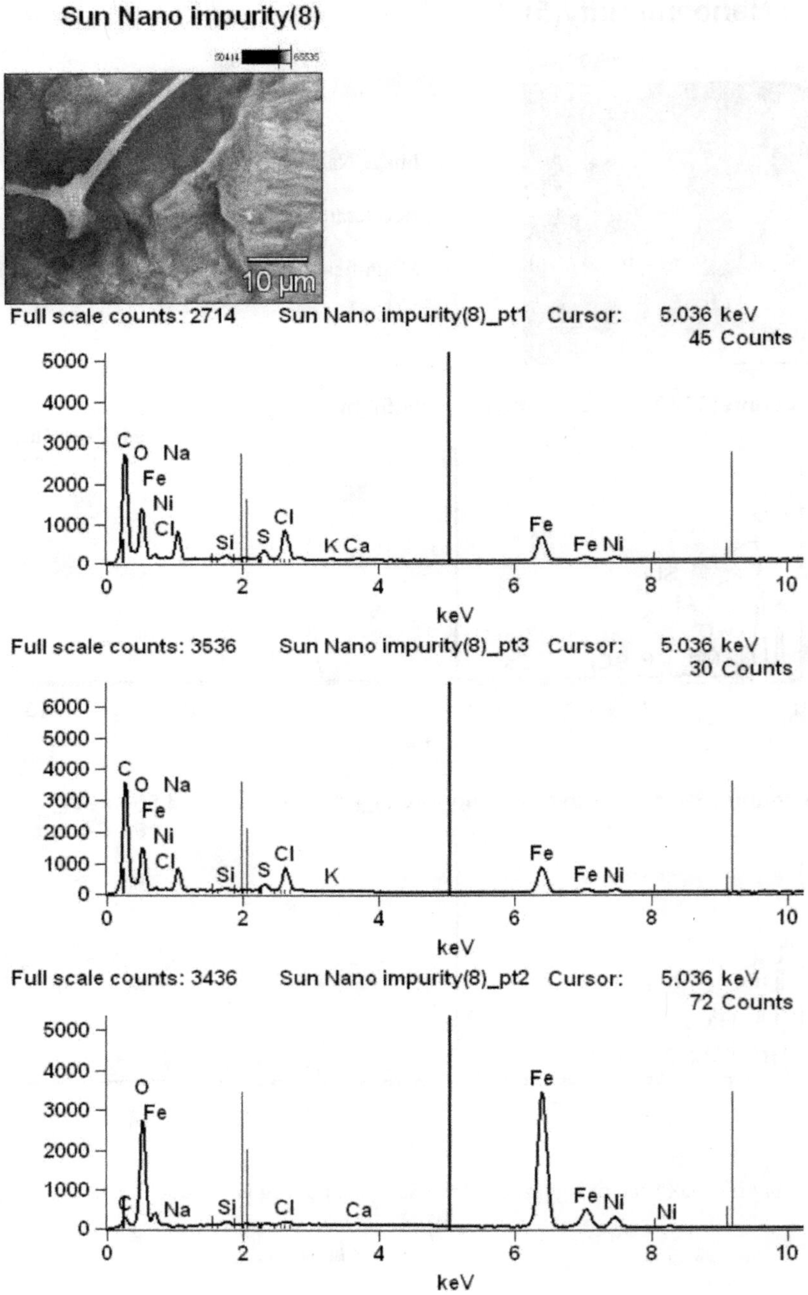
Sun Nano impurity(8)
10 µm
Full scale counts: 2714
Sun Nano impurity(8)_pt1
Cursor: 5.036 keV
45 Counts
C O Na Fe Ni Cl Si S Cl K Ca Fe Fe Ni
keV
Full scale counts: 3536
Sun Nano impurity(8)_pt3
Cursor: 5.036 keV
30 Counts
C O Na Fe Ni Cl Si S Cl K Fe Fe Ni
keV
Full scale counts: 3436
Sun Nano impurity(8)_pt2
Cursor: 5.036 keV
72 Counts
O Fe C Na Si Cl Ca Fe Fe Ni Ni
keV

Figure 42: Small Area EDX Scan of Light Area of the Sample Showing Bone-like Structure

APPENDIX D | 273

Sun Nano impurity(6)

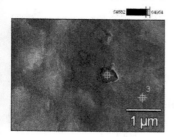

Image Name: Sun Nano impurity(6)

Accelerating Voltage: 20.0 kV

Magnification: 27000

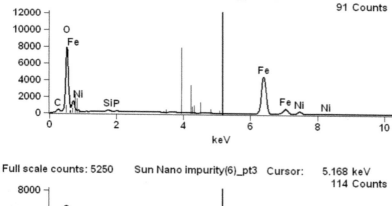

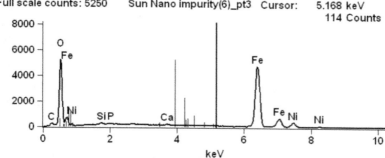

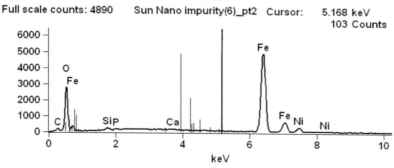

Figure 43: Small Area EDX Scan of Dark Area of the Sample Showing Nanopit in Dark Material

Sun Nano impurity(11)

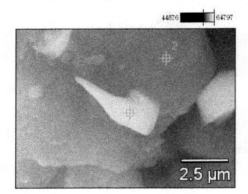

Image Name: Sun Nano impurity(11)

Accelerating Voltage: 25.0 kV

Magnification: 10000

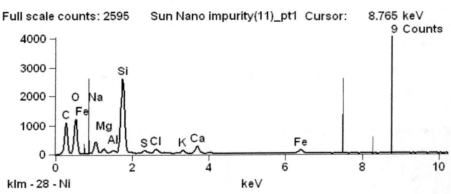

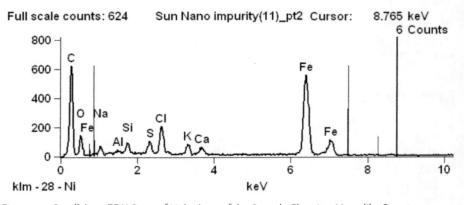

Figure 44: Small Area EDX Scan of Light Area of the Sample Showing Horn-like Structure

APPENDIX D | 275

Sun Nano impurity(12)

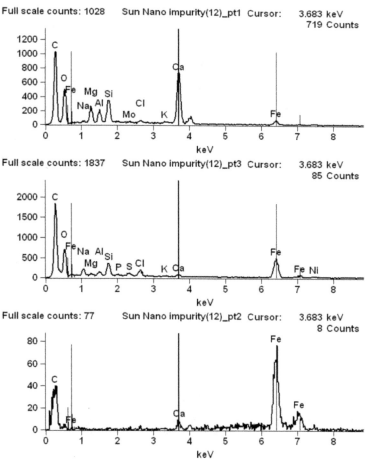

Image Name:
Sun Nano impurity(12)

Accelerating Voltage: 20.0 kV

Magnification: 3000

Figure 45: Small Area EDX Scan of Adjacent Light and Dark Areas of the Sample

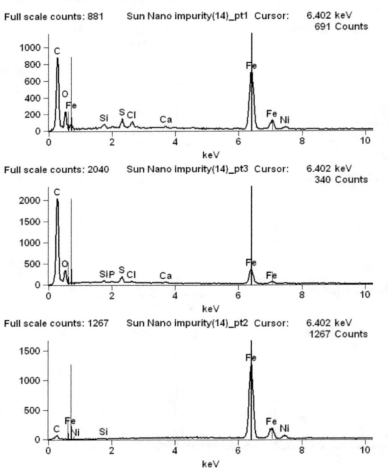

Figure 46: Small Area EDX Scan of Crack in Light Area of the Sample Showing Nanotubes

APPENDIX D | 277

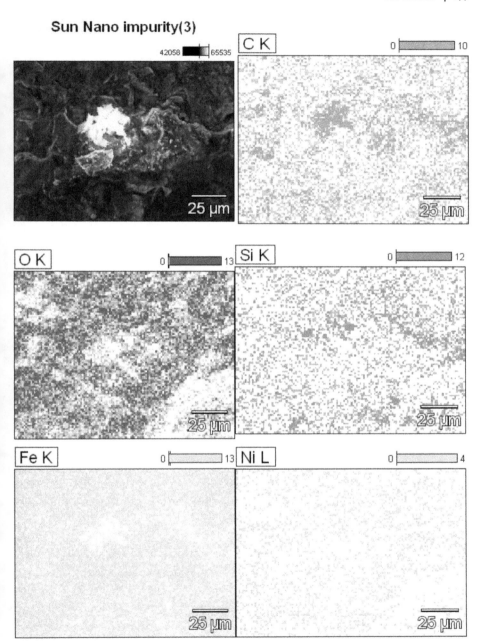

Figure 47: EDX Elemental Mapping of Sample Showing less Fe and Ni, and more C, O, and Si in Light Colored Inclusion

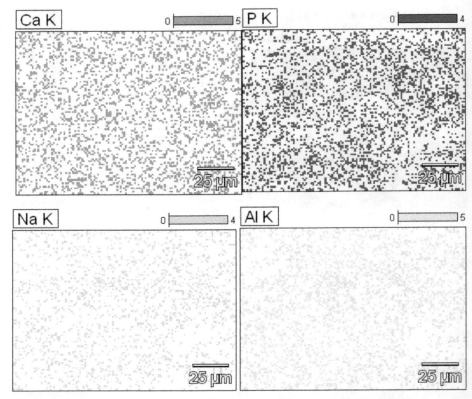

Figure 47 (continued): EDX Elemental Mapping of Sample Showing more Na and Al in Light Colored Inclusion

An unusual, horn-like, structure was observed (Figure 44), which showed an unusually high concentration of silicon, as compared to the other areas of light material which were imaged.

Small (~500 nm) orthorhombic crystals were also imaged (Figure 48), which were mainly composed of sodium and chlorine, with smaller amounts of sulfur, potassium, and iron. It is likely that the metallic elements in the crystals are present as the chlorides, and sulfates. These crystals were distributed in many locations on the dark material of the sample, and were uniform in size and shape.

EDX elemental mapping of larger areas of the sample (Figure 47), confirms that the composition of the light and dark areas are significantly different, with much more carbon, oxygen, silicon, sodium, and aluminum present in the

Sun Nano impurity(10)

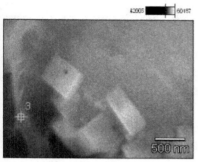

Image Name: Sun Nano impurity(10)

Accelerating Voltage: 25.0 kV

Magnification: 37000

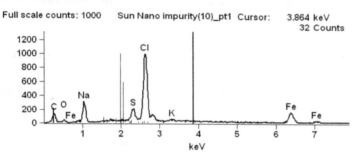

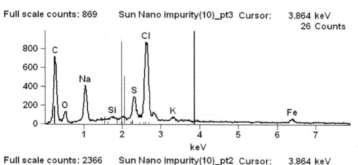

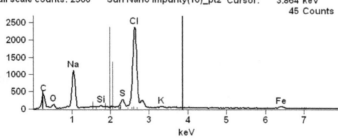

Figure 48: Small Area EDX Scan of Crystals and Adjacent Areas of Sample

lighter areas, along with less iron and nickel. Calcium and phosphorus appear to be more uniformly distributed in the sample.

Magnetic Analysis
The sample was found to be strongly attracted to a neodymium-iron-boron (NIB) magnet.

Raman Spectroscopy
The results of the Raman analyses, for 532 nm and 633 nm excitation wavelengths, are shown in Figures 49–51. Raman spectra of the sample are compared to Raman spectra of iron oxide, a sample of iron meteorite, and single-walled carbon nanotubes (SWCNT).

Five of the peaks seen in the low wave-number part of the 532 nm Raman spectrum of the sample (Figure 49, 159 cm^{-1} to 679 cm^{-1}) appear to be a close match to similar peaks seen in the 532 nm Raman spectrum of the sample of iron meteorite.[5]

The first and last of these four peaks also appear to roughly match two of the peaks seen in the 532 nm Raman spectrum of the iron oxide sample.

Three very interesting peaks are seen in the high wave-number portion of the 532 nm Raman spectrum of the sample (1312.0 cm^{-1}, 1566.8 cm^{-1}, and 1587.7 cm^{-1}, Figure 50).

These three peaks appear to match the D-band, and both metallic (Met) and semiconducting (SC) G-band peaks of the 532 nm Raman spectrum of the single-walled carbon nanotube sample. The sample peaks appear up-shifted by approximately 3 cm^{-1}–4 cm^{-1}, relative to the carbon nanotube sample selected for comparison.

If these peaks are produced by the presence of carbon nanotubes, it is likely that the 159 cm^{-1} peak is a carbon nanotube radial breathing mode (RBM) band peak.

---

5   This was a sample of the Campo de Cielo iron meteorite, from Argentina.

APPENDIX D | 281

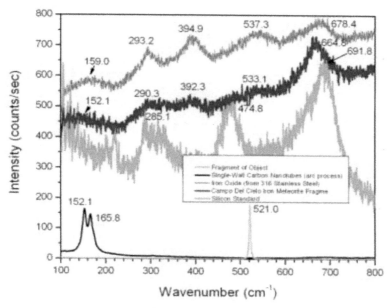

Figure 49: 532 nm Raman Spectrum of Sample as compared to iron oxide, a sample of iron meteorite, and single-walled carbon nanotubes – low Wave-Numbers

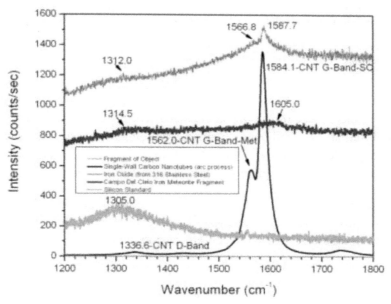

Figure 50: 532 nm Raman Spectrum of Sample as compared to iron oxide, a sample of iron meteorite, and single-walled carbon nanotubes – high Wave-Numbers

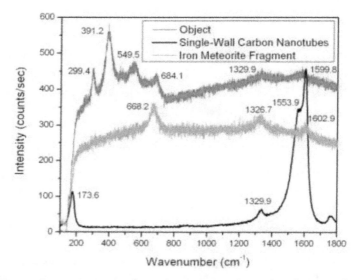

Figure 51: 633 nm Raman Spectrum of Sample as compared to iron oxide, a sample of iron meteorite, and single-walled carbon nanotubes – full Spectrum

The 633 nm Raman spectrum of the sample (Figure 51) shows what appear to be the same peaks seen at 532 nm, with somewhat greater intensities for the peaks in the lower wave-number portion of the spectrum. The peak corresponding to the CNT G-Band is somewhat less intense, however.

ICP-MS Analysis

An inductively-coupled plasma-mass spectroscopy (ICP-MS) analysis was performed on a piece of the object removed from Mr. Smith, after the SEM, EDX, and Raman data had been obtained. The ICP-MS analysis was performed by an independent laboratory. The piece of the sample used was the same one which was analyzed in the previous tests in this report.

The sample piece was digested in a mixture of nitric and hydrochloric acids,[6] and an aliquot of the liquid analyzed by ICP-MS. A portion of the sample, which did not dissolve, proved to contain carbon nanotubes.[7]

---

6   See bottom of next page for details of the digestion process.
7   This was shown by Raman analysis of the residue; carbon nanotubes are very resistant to the action of most reactive chemicals, and would be expected to survive this type of acid digestion process.

The ICP-MS elemental analysis confirmed the EDX results, concerning the major components of the sample, and also found many trace elements, which were not detected by EDX (Table 1). A total of fifty-one (51) elements were detected in the sample by ICP-MS.[8]

The major components of the sample, in order of abundance, were iron (> 46%),[9] and nickel (5.20%).

Minor component elements, in order of abundance (Table 2), were silicon (0.27%), cobalt (0.22%), phosphorus (0.16%), and calcium (0.15%).

Table 1 – Results of ICP-MS Analysis of Piece of Implant Sample

Trace Impurities by SOP 7040, Rev 9
Inductively Coupled Plasma – Mass Spectrometry

	ppm	Detection Limit		ppm	Detection Limit
Aluminum	260	30	Molybdenum	9.3	0.05
Antimony	0.37	0.2	Neodymium	0.39	0.02
Arsenic	17	0.4	Nickel	52000	0.1
Barium	96	0.1	Niobium	0.37	0.1
Beryllium	ND	0.05	Osmium	0.37	0.1
Bismuth	ND	0.03	Palladium	3.3	0.02
Boron	15	3	Phosphorus	1600	10
Bromine	ND	5	Platinum	10	0.02
Cadmium	ND	0.09	Potassium	ND	50
Calcium	1500	30	Praseodymium	0.11	0.02
Cerium	0.85	0.03	Rhenium	0.66	0.02
Cesium	ND	0.02	Rhodium	2.8	0.02
Chromium	13	0.2	Rubidium	0.15	0.02
Cobalt	2200	0.09	Ruthenium	8.0	0.02
Copper	170	0.3	Samarium	0.13	0.02
Dysprosium	0.11	0.02	Selenium	2.5	1
Erbium	0.07	0.02	Silicon	2700	50

---

8 The ICP-MS analysis was not sensitive to halogens (F, Cl, Br, I), or sulfur (S).
9 An exact percentage of iron cannot be obtained from this analysis, because the mass spectrometer detector was saturated. The EDX-derived value for the percentage of iron in the sample (~94%) is more accurate, in this case.

	ppm	Detection Limit		ppm	Detection Limit
Europium	0.03	0.02	Silver	ND	0.02
Gadolinium	0.13	0.02	Sodium	230	10
Gallium	130	0.02	Strontium	10	0.2
Germanium	300	0.1	Tantalum	ND	0.07
Gold	0.90	0.09	Tellurium	ND	0.1
Hafnium	0.10	0.02	Thallium	ND	0.2
Holmium	ND	0.02	Thorium	0.23	0.02
Iodine	ND	0.9	Thulium	ND	0.02
Iridium	3.6	0.05	Tin	6.5	0.1
Iron	460000	4	Titanium	20	0.3
Lanthanum	ND	1	Tungsten	1.9	0.07
Lead	1.3	0.1	Uranium	0.21	0.02
Lithium	ND	1	Vanadium	21	1
Lutetium	ND	0.5	Ytterbium	0.05	0.02
Magnesium	890	5	Yttrium	0.88	0.4
Manganese	62	0.1	Zinc	44	2
Mercury	ND	0.1	Zirconium	4.4	0.3

Note: The entire sample (0.0058 g) was mixed with 0.5 mL nitric acid and 0.5 mL hydrochloric acid and heated on a hot block set at 110°C for 1 hour. The mixture was cooled, 0.5 mL 30% hydrogen peroxide added, and heated 30 min. A portion remained undissolved; therefore, silicon or other elements that have low solubility in this acid mixture may be biased low. The solution was mixed with internal standards, diluted to 10 g, and a 1:100 dilution also analyzed by ICPMS.
Date Analyzed: 11-26-08
Elements Not Analyzed:
All Gasses, C, S, Sc, In, Tb

Table 2 – Elements Detected in Implant Sample by ICP-MS in Order of Abundance

	ppm	Detection Limit		ppm	Detection Limit
Iron	> 460000	4	Gold	0.90	0.09
Nickel	52000	0.1	Yttrium	0.88	0.4
Silicon	2700	50	Cerium	0.85	0.03
Cobalt	2200	0.09	Rhenium	0.66	0.02
Phosphorus	1600	10	Neodymium	0.39	0.02
Calcium	1500	30	Niobium	0.37	0.1
Magnesium	890	5	Antimony	0.37	0.2
Germanium	300	0.1	Thorium	0.23	0.02
Aluminum	260	30	Uranium	0.21	0.02
Sodium	230	10	Rubidium	0.15	0.02
Copper	170	0.3	Samarium	0.13	0.02
Gallium	130	0.02	Gadolinium	0.13	0.02
Barium	96	0.1	Dysprosium	0.11	0.02
Manganese	62	0.1	Praseodymium	0.11	0.02
Zinc	44	2	Hafnium	0.10	0.02
Vanadium	21	1	Erbium	0.07	0.02
Titanium	20	0.3	Ytterbium	0.05	0.02
Arsenic	17	0.4	Europium	0.03	0.02
Boron	15	3			
Chromium	13	0.2			
Strontium	10	0.2			
Platinum	10	0.02			
Molybdenum	9.3	0.05			
Ruthenium	8.0	0.02			
Tin	6.5	0.1			
Zirconium	4.4	0.3			
Iridium	3.6	0.05			
Palladium	3.3	0.02			
Rhodium	2.8	0.02			
Selenium	2.5	1			
Osmium	2.2	0.09			
Tungsten	1.9	0.07			
Lead	1.3	0.1			

Major trace elements detected included magnesium (890 ppm, or 0.089%), germanium (300 ppm), aluminum (260 ppm), sodium (230 ppm), copper (170 ppm), and gallium (130 ppm).

Minor trace elements included boron (15 ppm), barium (96 ppm) and strontium (10 ppm), titanium (20 ppm), vanadium (21 ppm), chromium (13 ppm), manganese (62 ppm), zinc (44 ppm), and arsenic (17 ppm).

The sample also contained smaller amounts of precious metals, including platinum (10 ppm), ruthenium (8.0 ppm), iridium (3.6 ppm), palladium (3.3 ppm), rhodium (2.8 ppm), osmium (2.2 ppm), and gold (0.90 ppm).

Other transition elements present included molybdenum (9.3 ppm) and tungsten (1.9 ppm), tin (6.5 ppm), zirconium (4.4 ppm) and hafnium (0.10 ppm), yttrium (0.88 ppm), rhenium (0.66 ppm), and niobium (0.37 ppm).

The sample also contained traces of rare earth elements, and actinides. The rare earth elements detected included cerium (0.85 ppm), neodymium (0.39 ppm), samarium (0.13 ppm), gadolinium (0.13 ppm), dysprosium (0.11 ppm), praseodymium (0.11 ppm), erbium, (0.07 ppm), ytterbium (0.05 ppm), and europium (0.03 ppm). Actinides detected included thorium (0.23 ppm), and uranium (0.21 ppm).

The remaining trace elements detected included selenium (2.5 ppm), lead (1.3 ppm), antimony (0.37 ppm), and rubidium (0.15 ppm).

Potassium was not detected in the ICP-MS analysis, although it was detected in the sample by EDX. It is likely that the amount of potassium in the sample was below the detection limit of the ICP-MS analysis (50 ppm).

Isotopic Analysis
The raw ICP-MS data had sufficient resolution to calculate percentages of isotopes for four of the elements detected in the sample (boron, magnesium, nickel, and copper).

The distribution of isotopes in the elements the sample is made of is a strong indication of the area that the sample was formed. Any deviation of more than

1% of the isotopic ratios in the sample from the terrestrial isotopic distribution indicates that the sample was probably not formed on Earth.

The data showed significant differences between the isotopic distributions of most of the sample elements, for which isotopic data was available, and the isotopic distributions of the same elements obtained from Earthly sources.

The isotopic distributions of the elements in the sample differed by as much as 4% from the terrestrial distributions of the same elements, indicating that the sample probably did not originate on Earth.

Table 3 – Isotopic Abundances of Elements Detected in Implant Sample

Isotope	Sample Isotopic Abundance (%)	Terrestrial Isotopic Abundance (%)	Error (%)
Sb^{121}	49.68	57.36	3.6
Sb^{123}	50.32	42.64	5.7
B^{10}	54.24	19.9	12.2
B^{11}	45.76	80.1	9.0
Mg^{24}	49.61	78.99	1.2
Mg^{25}	50.39	10.00	1.2
Mg^{26}	50.39	11.01	0.2
Ni^{58}	29.13	68.08	NS
Ni^{60}	35.03	26.23	1.0
Ni^{61}	0	1.14	NS
Ni^{62}	35.84	3.63	1.0
Ni^{64}	0	0.93	NS
Cu^{63}	49.74	69.15	1.4
Cu^{65}	50.26	30.85	2.1

Discussion

The iron/nickel metal matrix which made up the majority of the sample which was analyzed bore a strong resemblance to an iron-nickel meteorite.

This is seen by the similarity of the light microscope images of the sample to those of an iron meteorite sample, by the traces of iridium and tungsten seen in the EDX analysis, and by the similarity of the Raman spectrum of the sample to that of a sample of the Campo del Cielo iron-nickel meteorite.

The resemblance of the sample to a meteorite was confirmed by the pattern of trace elements detected in the ICP-MS analysis. The analysis confirmed the presence of traces of iridium, which is very rare on earth, but is universally present in meteoric iron.

The analysis also showed the presence of relatively large amounts of gallium (130 ppm) and germanium (300 ppm), which are also generally present in iron-nickel meteorites, at concentrations of up to 100 ppm for gallium (Ga), and up to 400 ppm for germanium (Ge). These concentrations of Ga and Ge would therefore be considered to be at the high end of the concentration range for these elements, in an iron-nickel meteorite.

The presence of traces of precious metals, other than iridium, is also a good indicator of possible meteoric origin of the metallic portion of the sample. The elements carbon (C), copper (Cu), cobalt (Co), sulfur (S), phosphorus (P), chromium (Cr), gallium (Ga), germanium (Ge), arsenic (As), antimony (Sb), tungsten (W), rhenium (Re), iridium (Ir), gold (Au), ruthenium (Ru), palladium (Pd), osmium (Os), praseodymium (Pr), and manganese (Mn) have all been detected in iron-nickel meteorites. All of these elements were also detected in the ICP-MS analysis of the sample from Mr. Smith s toe.

If the sample matrix material is derived from meteoric iron, its nickel content (5.2%, by ICP-MS, 6% by EDX) is somewhat low, but within the range of nickel percentages published in the literature (5%–25%), for known iron-nickel meteorites. This concentration of nickel would place the material in a class of low-nickel iron meteorites, known as hexahedrites.

The differences in the isotopic ratios of the sample elements from those which are observed in elements derived from terrestrial sources were quite remarkable, and cannot be easily explained, except by an extraterrestrial origin of the sample material.

Isotopic percentages in elements derived from terrestrial sources have not been observed to vary by more than ±1%, at most, while the variations in the percentages of isotopes observed in many of the sample elements were more than 2× greater than this.

Some lighter isotopes of meteoric origin, such as boron, have been observed to vary in isotopic percentages by up to 19%, relative to the same elements derived from terrestrial sources. Heavier elements from known meteorites[10] have not been observed to vary in isotopic abundances, with respect to terrestrial standards, by nearly as much as was observed in all of the elements which were analyzed for isotopic distribution in the ICP-MS test of the current sample.

This could indicate that the material is not only extraterrestrial, but may originate from a different solar system than our own. The point of origin of the sample material may perhaps lie nearer the center of our galaxy, where supernovae[11] are more common.

The high percentage of iron observed in the chemical analysis data indicates strongly that the red patina on the sample, as delivered, was hydrated iron oxide (rust), a corrosion product formed by contact of the iron in the sample with oxygen, and the water present in the blood serum the sample was stored in. The salts dissolved in the blood serum undoubtedly accelerated the corrosion of the metallic portion of the sample.

The black material, seen on the sample pieces by Dr. Leir, soon after removal from the patient, was freshly formed iron oxide,[12] in which hydration had not yet been completed.

---

10 Isotopic analyses of iron meteorite trace elements were found in the literature for Ni, Pt, Pd, Os, and Rh.
11 Supernovae create heavy elements by the r (rapid) process of neutron-capture nucleosynthesis. Neutronrich isotopes of heavy elements are thought to form in supernova explosions.
12 This material (Fe_2O_3) is black in color, and turns red after long exposure to water.

The sample had an outer coating of a non-metallic, ceramic-like material, which was approximately 100 nm–200 nm in thickness. This material had a somewhat rough texture, as seen under SEM, with surface irregularities up to several microns in size.

Large numbers of inclusions of what appeared to be the same material,[13] which were typically several microns in size, were also seen in the metallic phase.

The high concentration of non-metallic inclusions in the metallic phase probably account for the brittleness of the original object. Inclusions of unlike material, in this size range, which do not bind well with the sample matrix, act as points of stress concentration during episodes of mechanical stress, which leads to cracking at much lower stress levels than would be the case with a homogeneous metallic material.

The presence of these inclusions is the most likely cause of the breakage of the original object into small pieces, during its removal from Mr. Smith's toe.

The non-metallic, ceramic-like, material contains mainly carbon (C), oxygen (O), silicon (Si), sulfur (S), aluminum (Al), calcium (Ca), iron (Fe) and nickel (Ni), with smaller amounts of sodium (Na), phosphorus (P), chlorine (Cl), potassium (K), and titanium (Ti), and chemically resembles a biological hard part, such as shell, or bone.[14]

This similarity of the composition of the non-metallic phase to biological material may be responsible for the lack of immune response to the object by the patient s body.

The opalescence of this material, seen in the light microscopy images, indicates the presence of an organized, layered, structure, such as occurs in mother-of-pearl, or opal, which reflects and refracts light strongly into different color bands.

The Raman data, showing what appears to be carbon nanotube D-Band and G-Band signals, along with a possible radial breathing mode signal, strongly

---

13 EDX elemental analyses of the inclusions and the outer coating were very similar.
14 Elemental composition of the non-metallic, ceramic-like phase of the sample was derived from EDX data.

indicates the presence of carbon nanotubes (CNTs). This is confirmed by the SEM images, which show bundles of nanotubes, with high carbon content (EDX data), which appear nearly identical to SEM images of commercial arc-process, single-walled, CNTs.

The data therefore indicates that the majority of the non-metallic phase material is probably composed mainly of carbon nanotubes, which are covered, and/or filled, by a shell-like coating of aluminum, calcium, iron, nickel, and titanium silicates, oxides, sulfates, and phosphates.

A smaller percentage of the non-metallic phase of the sample is composed of the very regularly sized (500 nm), and shaped, sodium, potassium, and iron, chloride and sulfatecontaining crystals, seen in the SEM images. These crystals appear to be far too regular in size and shape to have formed spontaneously, from drying of the salts in the blood serum the samples was stored in.

The shapes of the inclusions of the lighter, non-metallic, material in the Fe/Ni phase appear to be non-random, such as the long bone-like, and horn-like structures seen in the SEM images. The Fe/Ni phase also has numerous pits, of regular size (400 nm–500 nm) and shape.

The carbon nanotubes inside the above structures would be excellent carriers of electric current, and could also act as electronic components, depending on whether the CNT type were metallic, or semiconducting. The shell-like coating on the material would then provide good electrical insulation for these nano-components.

The relatively large amounts of silicon and germanium in the sample may also be indicative of the presence of silicon-based, and/or germanium-based electronic components in the sample.

It is not likely that inclusions of the type observed could have formed within molten iron/nickel, during the formation process of an iron-nickel meteorite, since the solubilities of most ceramics in molten iron is high enough to dissolve a small percentage of ceramic inclusions, especially if these were small in size. It is also difficult to conceive of a natural process in which ceramic inclusions of this type could be formed inside metal, while the metal is in the solid state.

There are, in any event, no known meteorites which contain ceramic inclusions of this type. This is an anomaly, considering the fact that all the other evidence appears to point to a meteoric origin for the sample.

Because of this observation, the observation that the inclusions appear to be artificially shaped nano-components, and the fact that the complete object was giving off radio signals, before removal, the conclusion is inescapable that the object the sample came from is a manufactured item, which was made using extraterrestrial materials, by an organization possessing a high degree of technological sophistication.

The differently shaped inclusions of the non-metallic, CNT-containing material could then be for the performance of different functions in the device, such as carrying electric current (bone-like structures), acting as antenna for emitting, or receiving, radio signals (horn-like structures), or acting as resonators, to generate the radio waves (salt-containing crystals).

The magnetic nature of the metallic matrix material of the object may be necessary for the object, or device, to function, or, alternatively, meteoric iron could have been chosen as a base material simply because of its abundance in our solar system, and probably other solar systems, as well.

This would make it a relatively inexpensive material for use in manufacture, if the organization, or society, that made the object already had inexpensive methods of space transport, and travel.

The manufacture of a device comparable to this one is probably beyond the technology of known, Earthly, commercial processes, at the present time. It is most likely, therefore, that the device was manufactured by an alien civilization. It is still a possibility, however, that the device was manufactured by some process known to the Earthly military/black Project community.

The hypothesis that this sample came from a manufactured, nanotechnological device should be investigated by further research on this, and all other such samples, obtained previously.

Further research should include include Raman analysis, to check for the presence of carbon nanotubes in the older samples, and their analysis by ICP-MS.

One, or more, samples should also be subjected to a combination of SEM imaging, EDX elemental analysis, and etching by fast-atom bombardment. This type of technique should allow a layer by layer, three dimensional, image to be built up of the composition of the sample. This type of information may be essential to understanding of the function of these devices. Study of the electrical characteristics of these samples may also be of great benefit.

Conclusions

1. The sample consists mainly of iron, with a high carbon and oxygen content. The iron base material contains 5.2% nickel, and is highly magnetic. Traces of iridium, and other precious metals, tungsten, gallium, and germanium present strongly suggest that the metallic portion of the sample was derived from meteoric iron, and is extraterrestrial in origin.

2. The extreme differences in the isotopic ratios of the sample elements from the isotopic ratios of elements found on Earth provide strong confirmation that the material in the sample is of extraterrestrial origin.

3. The sample consists of two major phases; an iron/nickel (Fe/Ni) phase, and a nonmetallic phase resembling a hard biological substance, such as shell, tooth, or bone. The iridescence of the non-metallic phase, seen in light microscopy, suggests a layered microstructure, perhaps similar to mother-of-pearl, or opal.

4. The similarity of the composition of the non-metallic phase to biological material may be responsible for the lack of immune response to the object by the patient s body.

5. This non-metallic phase is high in carbon, oxygen, silicon, magnesium, aluminum, sulfur, and phosphorus, and is present as an outer covering on the sample, and as inclusions in the metallic, Fe/Ni phase.

6. The non-metallic phase of the sample also contains bundles of carbon nanotubes, perhaps covered, or filled, with calcium and magnesium silicates, phosphates and sulfates.

7. The inclusions of the non-metallic phase have unusual shapes, which suggest artificiality, and functionality. This, along with the fact that the

object was giving off radio signals, before removal, strongly indicates that this is a manufactured, nanotechnological device, which was inserted in patient Smith for a definite purpose.

8. The function of the device cannot be determined with certainty from the available data, and the device may have had multiple functions and missions. It is likely, however, that two of its functions had to do with monitoring of the physiological state of Mr. Smith s body, and mood/mind control.

Bibliography

"Lithium Isotope Analysis of Inorganic Constituents of the Murchison Meteorite" Mark A. Sephton, et. al.; *The Astrophysical Journal*, 612:, September 1, 2004

"Boron Isotope Ratios in Meteorites and Lunar Rocks" Zhai, et. al., *Geochimica et Cosmochimica Acta*, vol. 60, Issue 23, 12/1996

"Carbon Isotope Abundance in Meteoritic Carbonates" Robert N. Clayton; *Science*, 12 April (1963)

"Ti 50 Anomalies in Primitive and Differentiated Meteorites" Goldschmidt Conference Abstracts, A1038 (2007)

"Measurements of the Isotopic Ratios of Nickel in Iron Meteorities, Using Nickel Carbonyl" Masaru Suzuki and Sadao Matsuo; *Geochemical Journal*, vol. 1, 1967

"Nickel Isotope Anomalies in Meteorites and the Fe 60–Ni 60 Clock" M. Bizzarro, et. al.; *Workshop on Chronology of Meteorites* (2007)

"Germanium Isotopic Fractionation in Iron Meteorites" 64th Annual Meteoritical Society Meeting (2001)

"Tungsten Isotope Evidence from 3.8–Gyr Metamorphosed Sediments for Early Meteorite Bombardment of the Earth" Shoenberg, et. al.; *Nature* 418, (25 July 2002)

"Tungsten Isotopic Constraints on the Formation and Evolution of Iron Meteorite Parent Bodies" A. Markowski, et. al.; Dept. of Earth Sciences, Oxford Univ.; *Lunar and Planetary Science* XXXVI (2005)

"Microanalysis of Platinum Group Elements in Iron Meteorites Using Laser Ablation ICP-MS" A.J. Campbell and M. Humayun, Dept. of Geophysical Sciences, Univ. of Chicago; *Lunar and Planetary Science* XXX (1974)

"New Applications of the Re187–Os187 and Pt190–Os186 Systems to the Study of Iron Meteorites" D.L. Cook, et. al., Isotope Geochemistry Laboratory, Dept. of Geology, Univ. of Maryland; *Lunar and Planetary Science* XXXI

"Compilation of Minimum and Maximum Isotope Ratios of Selected Elements in Naturally Occurring Terrestrial Materials and Reagents" Coplen, et. al.; U.S. Geological Survey Water Resources Investigation Report

"Meteorites: A Petrologic, Chemical and Isotopic Synthesis" *Cambridge Planetary Science Series* R. Hutchinson; Cambridge University Press (2004)

Bibliography

Books

Alexander, Col. John B.: *UFOs: Myths, Conspiracies, and Realities.* Thomas Dunne Books, St. Martin's Press, 2011

Andrews, Ann and Jean Ritchie: *Abducted; The True Story of Alien Abduction in Rural England.* Headline Book Publishing, London, 1998

Blumenthal, Ralph: *The Believer: Alien Encounters, Hard Science, and the Passion of John Mack.* High Road Books, University of New Mexico, 2021

Brachthauser, Christian: *Geheimnisvolle Grauzone, Eine Kritische Analyse des Abduktionsphaenomens.* Books on Demand GmbH, 2001 [NB no English language translation of this work is available but if you read German, it's recommended]

Breccia, Stefano: *Mass Contacts.* Author House, Milton Keynes, MK9 2BE, UK, 2009

Brener, Milton E.: *Walking through Walls and other Impossibilities: The Hybrid Agenda.* Xlibris Corporation, 2007

Bryan, C. D. B.: *Close Encounters of the Fourth Kind: Alien Abduction and UFOs – Witnesses and Scientists Report.* Orion Publishing, London, 1995

Bullard, Thomas E.: *The Myth and Mystery of UFOs.* University Press of Kansas, 2010

Cahill, Kelly: *Encounter.* Harper Collins, Australia, 1996

Chalker, Bill: *Hair of the Alien: DNA and other Forensic Evidence of Alien Abductions.* Paraview Pocket Books, a division of Simon and Schuster, New York, 2005

Collings, Beth and Anna Jamerson: *Connections: Solving our Alien Abduction Mystery.* Wild Flower Press, Newburg OR, 1996

Dolan, Richard: *UFOs and the National Security State: Chronology of a Cover-up, 1941-73.* Keyhole Publishing/Hampton Roads, 2002

Dolan, Richard: *UFOs and the National Security State: The Cover-up Exposed, 1973-91.* Keyhole Publishing, Rochester NY, 2009

Donderi, Don: *UFOs, ETs and Alien Abductions; A Scientist Looks at The Evidence.* Hampton Roads, Charlottesville, VA, 2013

Donderi, Don: *Truth, Lies and ETs: How we stumbled into the Universe.* Abeville, SC: Moonshine Cove, 2022

Druffel, Ann and D. Scott Rogo: *The Tujunga Canyon Contacts: A Continuing "Chain Reaction" of UFO Encounters and Abductions.* Prentice Hall, 1980

Fielding, Peggy: *The Story of Barbara Bartholic, UFO Investigator.* AWOC, Denton TX, 2004

Fiore, Edith: *Abductions; Encounters with Extraterrestrials.* Doubleday Bantam Bell Publishing Group, New York, 1989

Fowler, Raymond E: *The Andreasson Affair: The Documented Investigation of a Woman's Abduction Aboard a UFO.* Prentice-Hall International, 1979

Fowler, Raymond E.: *The Watchers: The Secret Design Behind UFO Abduction.* Bantam Books, New York, 1982

Fowler, Raymond E.: *The Watchers II: Exploring UFOs and the Near-Death Experience.* Wild Flower Press, Newburg OR, 1995

Fowler, Raymond E.: *The Andreasson Legacy: UFOs and The Paranormal.* Marlowe and Company, New York, 1997

Fowler, Raymond E.: *The Allagash Abductions: Undeniable Evidence of Alien Intervention.* Wild Flower Press, Tigard OR, 1993

Fowler, Raymond E.: *UFO Testament: Anatomy of an Abductee*. Writer's Showcase, Lincoln NE, 2002

Fowler, Raymond E. and Luca, Betty Ann: *The Andreasson Affair Phase Two: The Continuing Investigation of a Woman's Abduction by Alien Beings*. Prentice-Hall International, 1982

Friedman, Stanton T. and Kathleen Marden: *Captured: The Betty and Barney Hill Story*. Career Press, Franklin Lakes NJ, 2007

Friedman, Stanton T: *Flying Saucers and Science: A Scientist Investigates the Mysteries of UFOs*. Career Press, Franklin Lakes, NJ, 2008

Fuller, John G.: *The Interrupted Journey: Two Lost Hours Aboard a Flying Saucer*. The Dial Press, New York, 1966

Genova, Lisa: *Remember: the Science of Memory and the Art of Forgetting*. Harmony Books, Penguin Random House LLC, New York, 2021

Good, Timothy: *Above Top Secret; The Worldwide UFO Cover-up*. Guild Publishing, by arrangement with Sidgwick and Jackson, 1987

Good, Timothy: *Alien Liaison: The Ultimate Secret*. Century Random House, London, 1991 [NB In the USA, this work was published as *Alien Contact*]

Good, Timothy: *Beyond Top Secret: The Worldwide UFO Security Threat*. Sidgwick and Jackson, London, 1996

Good, Timothy: *Alien Base: Earth's Encounters with Extraterrestrials*. Century Random House, London, 1998

Good, Timothy: *Unearthly Disclosure*. Century Random House, London, 2000

Good, Timothy: *Need to Know: UFOs, the Military and Intelligence*. Sidgwick and Jackson, 2006

Hansen, Terry: *The Missing Times: News Media Complicity in the UFO Cover-up*. Xlibris, 2000

Harpur, Patrick: *Daimonic Reality: A Field Guide to The Otherworld*. Pinewinds Press, 1994

Hastings, Robert: *UFOS and Nukes: Extraordinary Encounters at Nuclear Weapons Sites.* Author House, 2008

Hastings, Robert and Dr. Bob Jacobs: *Confession: Our Hidden Alien Encounters.* Kindle Direct Publishing, 2019

Hawking, Stephen: *A Brief History of Time, from the Big Bang to Black Holes.* Guild Publishing, 1992

Hawking, Stephen: *A Brief History of Time: A Reader's Companion.* Transworld Publishers, 1992

Hawking, Stephen: *The Universe in a Nutshell.* Transworld Publishers, 2001

Hickson, Charles and William Mendez: *UFO Contact at Pascagoula*, Wendelle Stevens Publishing, 1983

Hopkins, Budd: *Missing Time: A Documented Study of UFO Abductions.* Richard Marek Publishers, New York, 1981

Hopkins, Budd: *Intruders: The Incredible Visitations at Copley Woods.* Random House, New York, 1987

Hopkins, Budd: *Witnessed: The True Story of the Brooklyn Bridge UFO Abductions.* Pocket Books, New York, 1996

Hopkins, Budd: *Art, Life and UFOs.* Anomalist Books, New York, 2009

Hopkins, Budd and Carol Rainey: *Sight Unseen: Science, UFO Invisibility and Transgenic Beings.* Atria Books, New York, 2003

Hynek, J. Allen: *The UFO Experience, A Scientific Enquiry: A Critical Appraisal of the UFO Problem and its Investigation by the Foremost Authority Involved in this Research.* Abelard Schuman, New York and London, 1972

Hynek, J. Allen: *The Hynek UFO Report on Project Blue Book.* Barnes and Noble, 1977

Hynek, J. Allen and Jacques Vallee: *The Edge of Reality; A Progress Report on Unidentified Flying Objects.* Henry Regency, New York, 1975

Jacobs, David M.: *The UFO Controversy in America*. Indiana University Press, 1975

Jacobs, David M.: *Secret Life: Firsthand accounts of UFO Abductions*. Simon and Schuster, New York, 1992

Jacobs, David M.: *The Threat. The Secret Agenda: What the Aliens Really Want and How they Plan to Get it*. Simon and Schuster, New York, 1998

Jacobs, David M.: *Walking Among Us: The Alien Plan to Control Humanity*. Disinformation Books, San Francisco, 2015

Jacobs, David M., Editor: *UFOs and Abductions: Challenging the Borders of Knowledge*. University Press of Kansas, 2000

Jordan, Debbie and Kathy Mitchell: *Abducted! The Story of the Intruders continues*. Carroll and Graf, New York, 1994

Jordan-Kauble, Debrah: *Extraordinary Contact: Life Beyond Intruders*. White Crow Productions, 2021

Kaku, Michio: *Hyperspace: A Scientific Odyssey through Parallel Universes, Time Warps, and the 10th Dimension*. Oxford University Press, New York 1994

Kaku, Michio: *Parallel Worlds: The Science of Alternative Universes and Our Future in the Cosmos*. Penguin Group Publishing, 2005

Kean, Leslie: *UFOs: Generals, Pilots and Government Officials Go on the Record*. Harmony Books, Crown Publishing Group, Random House, New York, 2010

Kean, Leslie: *Surviving Death: A Journalist Investigates Evidence for an Afterlife*. Crown Archetype, Penguin Random House, New York, 2017

Keel, John A.: *Operation Trojan Horse: A Breakthrough Study in UFOs*. GP Puttnam, 1970

Kelleher, Colm A. and George Knapp: *Hunt for the Skinwalker*. Pocket Books (Simon & Schuster) New York, London, Toronto, Sydney, 2005

Kinsella, Philip: *Terrestrial Trespassers: The Greys, Abductions and Areas of High Strangeness*. Flying Disk Press, Pontefract UK, WF8 4QX, 2023

Lacatski, James, with Colm A. Kelleher and George Knapp: *Skinwalkers at the Pentagon: An Insiders' Account of the Secret Government UFO Program*. RTMA LLC, Henderson NV, 2021

Lamb, Barbara and Nadine Lalich: *Alien Experiences: 25 cases of Close Encounter Never Before Revealed*. Trafford Publishing, 2008

Leir, Roger K.: *The Aliens and the Scalpel: Scientific Proof of Extraterrestrial Implants in Humans*. Granite Publishing, Columbus NC, for The National Institute of Discovery Science, 2005

Lindemann, Michael (editor): *UFOs and the Alien Presence – Six Viewpoints*. The 2020 Group, Santa Barbara, CA, 1991

Lorenzen, Coral and Jim (for the Aerial Phenomena Research Organization): *Abduction: Confrontation with Beings from Outer Space*. Berkley Publishing, New York, 1977

Lovelace, Terry: *Incident at Devils Den: Compelling Proof of Alien Visitation, Alleged Government Involvement and an Alien Implant discovered on a routine X-Ray*. Published by Amazon, 2018.

Lovelace, Terry: *The Reckoning*. Published by Amazon, 2020.

MacGregor, Trish and Rob MacGregor: *Aliens in the Backyard: UFOs, Abductions and Synchronicity*. Crossroad Press, 2013

Mack, John E.: *Abduction: Human Encounters with Aliens*. Simon and Schuster, 1994

Mack, John E.: *Passport to the Cosmos: Human Transformation and Alien Encounters, Commemorative Edition*. White Crow Books, London, 2011 (new material copyright 2008)

Marden, Kathleen and Denise Stoner: *The Alien Abduction Files: The Most Startling Cases of Human-Alien Contact Ever Reported*. Career Press, Pompton Plains, NJ, 2013

Marden, Kathleen: *Forbidden Knowledge*. Self-published, Amazon 2022

Mitchell, Dr. Edgar: *The Way of the Explorer: An Apollo Astronaut's Journey Through the Material and Mystical Worlds*. Benito Lynch, Final Edition 2001

Miller, Dr. John G.: *Medical Procedural Differences: Alien Versus Human. Proceedings of the Abduction Study Conference, MIT, Cambridge, MA, 1992.* North Cambridge Press, 1994

Moulton Howe, Linda: *An Alien Harvest: Further Evidence Linking Animal Mutilations and Human Abductions to Alien Life Forms.* Linda Moulton Howe Productions, Jamison PA, Sixth Printing 2000

Mullis, Kary: *Dancing Naked in the Mind Field.* Random House, New York, 1998

Nagaitis, Carl and Philip Mantle: *Without Consent: A comprehensive Survey of Missing Time and Abduction Phenomena in the UK.* Ringpull Press, Cheshire, 1994

Pope, Nick: *Open Skies, Closed Minds: The First Time a Government UFO Expert Speaks Out.* Simon and Schuster, London, 1996

Pope, Nick: *The Uninvited: An Expose of the Alien Abduction Phenomenon.* Simon and Schuster, London, 1997

Pritchard, et al (Ed): *Alien Discussions: The Unabridged Proceedings of the Abduction Study Conference Held at MIT, Cambridge, MA.* North Cambridge Press, Cambridge MA, 1994

Randles, Jenny: *Abduction: Over 200 Documented UFO Kidnappings Exhaustively Investigated.* Robert Hale Publishing, 1988

Saito, Riichiro, Gene Dresselhaus, and Mildred S. Dresselhaus: *Physical Properties of Carbon Nanotubes.* London, Imperial College Press, 1998

Smith, Yvonne: *Chosen: Recollections of Abduction through Hypnotherapy.* Backstage Entertainment, Harbor City CA, 2008

Smith, Yvonne: *Coronado: The President, The Secret Service and Alien Abductions.* Amazon Publishing, 2014

Strieber, Whitley: *Communion: A True Story; Encounters with the Unknown.* Century Hutchinson, London and New York, 1987

Strieber, Whitley: *Transformation: The Breakthrough.* Century Hutchinson, London and New York, 1988

Strieber, Whitley: *Them.* Walker & Collier, San Antonio 78258 TX, 2023

Swann, Ingo: *Penetration: The Question of Extraterrestrial and Human Telepathy.* Swann-Ryder Productions, LLC; Illustrated edition, 2018 (originally pub 1998 by Swann Books)

Talbot, Michael: *The Holographic Universe.* Grafton Books/Harper Collins, 1991

Thompson-Smith, Angela: *Diary of an Abduction: A Scientist Probes the Enigma of her Alien Contact.* Hampton Roads, Charlottesville, VA, 2001

Turner, Dr. Karla and Ted Rice: *A Masquerade of Angels.* Kelt Works, Roland AK, 1994

Turner, Dr. Karla: *Into the Fringe: A True Story of Alien Abduction.* Berkley, New York, 1992 (2nd ed 2014)

Turner, Dr. Karla: *Taken: Inside the Alien-Human Abduction Agenda.* Kelt Works, Roland AK, 1994 (2nd ed 2014)

Vallee, Jacques: *Anatomy of a Phenomenon: Unidentified Objects in Space – A Scientific Appraisal.* Henry Regency, 1965

Vallee, Jacques: *Passport to Magonia: From Folklore to Flying Saucers.* Neville Spearman, 1970

Vallee, Jacques: *The Invisible College: What a Group of Scientists has Discovered about UFO Influences on the Human Race.* EP Dutton, New York, 1975

Vallee, Jacques: *Messengers of Deception: UFO Contacts and Cults.* And/Or Press, Berkeley CA, 1979

Vallee, Jacques: *Dimensions: A Casebook of Alien Contact.* Contemporary Books/UK publisher Souvenir Press, London, 1988

Vallee, Jacques: *Confrontations: A Scientist's Search for Alien Contact.* Ballantine Books, New York, 1990

Vallee, Jacques: *Revelations: Alien Contact and Human Deception.* Ballantine, New York, 1991

Vallee, Jacques: *Forbidden Science Volume One: Journals 1957-69: A Passion for Discovery.* Documatica Research, LLC, 2007

Vallee, Jacques: *Forbidden Science Volume Two: Journals 1970-79: California Hermetica*. Documatica Research, LLC, 2008 rev.2010

Vallee, Jacques: *Forbidden Science Volume Three: Journals 1980-89: On the Trail of Hidden Truths*. Documatica Research, LLC, 2012

Vallee, Jacques: *Forbidden Science Volume Four: Journals 1990-99: The Spring Hill Chronicles*. Lulu.com, 2017

Vallee, Jacques and Janine Vallee: *Challenge to Science*. Neville Spearman, 1966

Vallee, Jacques and Martine Castello: *UFO Chronicles of the Soviet Union: A Cosmic Samizdat*. Ballantine, New York,1992

Vallee, Jacques and Chris Aubeck: *Wonders in the Sky: Unexplained Aerial Objects from Antiquity to Modern Times*. Jeremy P. Tarcher/Penguin, 2010

Walton, Travis: *Fire in the Sky: The Walton Experience*. Skyfire Productions, Snowflake AZ, 2010 ed

Yurdozu, Farah: *Confessions of a Turkish Ufologist*. The Seven Houses, 2007

Other Sources

Bloecher, Ted, Budd Hopkins, and Jerry Stoehrer: *The Stonehenge Incidents – January 1975*. April 25, 1976. Prepared for presentation at the CUFOS Conference Lincolnwood, Illinois – April 30 to May 2, 1976. http://www.cufos.org/Abductions/The_Stonehenge_Incidents.pdf

Blumenthal, Ralph: *Alien Nation: Have Humans Been Abducted by Extraterrestrials?* VanityFair,May10,2013.https://www.vanityfair.com/culture/2013/05/americans-alien-abduction-science

Fox, Margalit: *Budd Hopkins, Abstract Expressionist and U.F.O. Author, Dies at 80*. New York Times, August 24, 2011. https://www.nytimes.com/2011/08/25/arts/design/budd-hopkins-abstract-artist-and-ufo-author-dies-at-80.html

Hopkins, Budd: *Sane Citizen Sees UFO in New Jersey*. Village Voice, March 1, 1976, p.12. https://alienjigsaw.com/et-contact/Sane%20Citizen%20Sees%20UFO%20in%20New%20Jersey.html

Notes and References

Introduction

1. There were the famous July 1952 flyovers of Washington DC by a number of UFOs, but they neither attempted to communicate nor land, let alone 'on the White House lawn'. https://en.wikipedia.org/wiki/1952_Washington,_D.C.,_UFO_incident https://www.youtube.com/watch?v=iRSkSQ3rFBw

2. On 24 June 1947, the mass media in the USA—and by extension, throughout the world—finally had a name for the strange flying objects pilots and other members of the population had for years past been seeing in the sky, after private pilot Kenneth Arnold reported a sighting of nine flying objects in a chain-formation 25 miles from Mount Rainier, Washington State in the Pacific Northwest. Arnold estimated the speed of the aerial craft to be around 1,700 miles per hour (faster than any man-made aircraft in the world at that time) and their movement pattern as "like saucers skipping on water." Although Arnold did not describe the objects as *saucer-shaped*, the name *flying saucers* was on that day born into popular culture and has remained entrenched there ever since.

3. Helen Cooper, Ralph Blumenthal, and Leslie Kean, "Glowing Auras and 'Black Money': The Pentagon's Mysterious U.F.O. Program," *New York Times*. December 16, 2017, https://www.nytimes.com/2017/12/16/us/politics/pentagon-program-ufo-harry-reid.html; Zachary Cohen, "What we know about UFOs: How the Pentagon has handled reported sightings, mysterious videos and more," CNN, May 17, 2021, https://edition.cnn.com/2021/05/17/politics/ufo-pentagon-explainer/index.html

4. See Chapter Seven, where my work with Dr David Jacobs over several years is recounted and the effects of hypnosis in enabling access to the long-term memory are explored in some detail

5. See chapters Three, Five, Eight, and Nine.

Prologue

1. Bees of Chester was sold in 1985; the name disappeared from the rose-growing world after being in business for over 100 years.
2. Steiger, *The Flying Saucer Menace*. "Rare 1967 UFO Magazine," https://rense.com/FSM/p1.htm
3. It looked exactly like the craft described by Bob Lazar. This is the principle reason why I find his claims credible: because as soon as I saw these images of what he called "the sport model," I instantly recognised this craft. It's not an invention. In 1972, I thought those black rectangles on the top part might have been darkened windows, but it seems they might have been something else. The one I saw at close range in 1972 even had the pointy 'antenna' thing on the top, just like that represented in this image: https://img1.cgtrader.com/items/2803254/30a60a2881/bob-lazar-sport-model-flying-saucer-3d-model-blend.jpg
4. Competent hypnosis by a skilled practitioner can break this barrier and enable access to the long-term memory. See Chapter Seven.

Chapter One

1. *Passport to Magonia*, by Jacques Vallée, was published in 1969. The ideas expounded in this book, how they often serve to confuse and inadvertently prevent understanding of the abduction program, are discussed in Chapter Five.
2. Steiger, *The Flying Saucer Menace*.
3. Both the Villas-Boas and the Hill cases are described in Chapter Two.
4. Moulton-Howe. *An Alien Harvest*. Howe's congressional testimony about her research into the cattle mutilations: https://www.youtube.com/watch?v=62NmS-R0WQo
5. "Discover Hurst Castle," https://www.hurstcastle.co.uk/ "Hurst Castle," English Heritage, https://www.english-heritage.org.uk/visit/places/hurst-castle
6. Strieber, *Communion*, https://en.wikipedia.org/wiki/Communion_(book)
7. "Bosa (Sardinia)," *Italy Magazine*, https://www.italymagazine.com/bosa
8. Explored in the 'Mindscan' section discussed in detail in Chapter Seven.
9. "Biography – Nick Pope," https://nickpope.net/wpte19/biography/
10. Hastings, *UFOs and Nukes*. Hastings and Jacobs, *Confession*. https://www.ufohastings.com/

11. "Taos, New Mexico – Visit Taos and Discover Northern New Mexico," https://taos.org/
12. See Chapter Four.
13. See Chapter Four.
14. See Chapter Eight.
15. The Spiritualist Association of Great Britain, https://sagb.org.uk/
16. Fowler and Luca, *The Andreasson Affair, Phase Two*.

Chapter Two

1. A 4-hour, extended TV miniseries titled, *Intruders*, based on the Debbie Jordan case, was also televised following the cinematic release.
2. SBESDV *Bulletin*, April–June 1962.
3. The program is intergenerational and runs in families. See Chapters One and Eight.
4. Further evidence that Barney Hill was not a regular abductee, in the program due to his ancestral line: had he been abducted previously, the abductors would have known this detail and not been surprised by it.
5. Fuller, *The Interrupted Journey*; Friedman and Marden, *Captured*.
6. Regarding the intergenerational aspect of the program, see Chapter Eight.
7. Fowler and Luca, *The Andreasson Affair, Phase Two*; Fowler, *The Watchers*; Fowler, *The Watchers II*; Fowler, *The Andreasson Legacy*.
8. Turner and Rice, *A Masquerade of Angels*.
9. Brian Broom, "'The story is very true. That's what has bothered me for 45 years.' UFO witnesses speak," *Clarion Ledger*, March 14, 2019, https://www.clarionledger.com/story/magnolia/2019/03/14/ufo-pascagoula-mississippi-calvin-parker-charles-hickson-other-witnesses/3129121002/
10. Jessica Potila, "Subject of 1976 UFO incident casts doubt on 'Allagash Abductions,'" *Fiddlehead Focus*, September 10, 2016, https://fiddleheadfocus.com/2016/09/10/news/community/top-stories/subject-of-1976-ufo-incident-casts-doubt-on-allagash-abductions/
11. Jordan and Mitchell, *Abducted*.

Chapter Three

1. Jacobs, *Secret Life*, pp. 91–96, contains a more complete account of the physical/medical examination endured by abductees. Other sources in the literature include, Pritchard et al., eds., *Alien Discussions*, which contains several such summaries among the papers and presentations. See the bibliography.
2. See Chapter Nine.
3. Hopkins, 'Note to the Reader,' in *Intruders*.
4. Ibid.
5. "We don't need no stinkin' badges!"https://en.wikipedia.org/wiki/Stinking_badges
6. Jacobs, *Secret Life*.
7. Randall Nickerson's documentary on the Ariel School Incident, *Ariel Phenomenon*, was released in 2022. https://arielphenomenon.com/
8. Hopkins, *Intruders*, p. 11, describes such an incident in detail, as reported by Debbie Jordan's elder sister, and speculates about the likely frequency of such episodes. You will have read about another case in the Prologue of this book.

Chapter Four

1. Peter Robbins and I first met at a MUFON conference in February 2008, as detailed in Chapter Six. We have since become close personal friends. http://peterrobbinsny.com/index.php/peter-s-bio
2. I was interviewed in December 2008 by Ricki Stern, the co-owner (with Annie Sundberg) of Breakthru Films, who was director of the now-aborted project. https://www.breakthrufilms.org/who-we-are
3. The 4-storey Chelsea property that Budd Hopkins owned and occupied for more than 50 years was mid-way between Union Square and the Meatpacking District, wherein lies the Whitney Museum of American Art. One can comfortably walk from the 16th Street location to Lower Manhattan, Wall Street, and the former site of the World Trade Center. The building at 246 w16th Street no longer exists, replaced in 2019 by a new building named The Grid, filled with expensive and upmarket apartments appropriate for this prime real estate location. https://streeteasy.com/building/the-grid
4. Leir. *The Aliens and the Scalpel*. First published in 1998 by The National Institute for Discovery Science, it has seen a couple of updated editions in the

current century, so there is more than one version available. The book runs to some 230 pages, including several detailed appendices. The first edition has a foreword penned by Whitley Strieber, following which the author introduces himself, his surgical credentials, and his history of interest in the subject of UFOs, abductees, and implants.

5. "Derrel W. Sims - Biography," Internet Movie Database, https://www.imdb.com/name/nm3356857/bio

6. Saito et al., *Physical Properties of Carbon Nanotubes*.

7. Steve Colbern's full report: https://docplayer.net/76207851-Analysis-of-object-taken-from-patient-john-smith-report-author-steve-colbern-25-january-2009.html

8. Marcus Lowth, "The Big Sur UFO Film Incident," *UFO Insight*, Last revised October 13, 2021, https://www.ufoinsight.com/ufos/cover-ups/big-sur-ufo-film; Robert Hastings, "UFOs Are Stalking and Intercepting Dummy Nuclear Warheads During Test Flights," August 23, 2011, https://ufohastings.com/articles/ufos-are-stalking-and-intercepting-dummy-nuclear-warheads-during-test-flights

9. Miller, "Medical Procedural Differences".

10. Ibid.

11. Ibid.

12. Chalker, *Hair of the Alien*.

13. Ibid.

14. Ibid.

15. Ibid.

16. Ibid.

17. Ibid.

18. Homozygous, as related to genetics, refers to having inherited the same versions (alleles) of a genomic marker from each biological parent. Thus, an individual who is homozygous for a genomic marker has two identical versions of that marker. By contrast, an individual who is heterozygous for a marker has two different versions of that marker. The PCR tests on Khuory's sample *implied* that the owner might be homozygous for the CCR5 Delta-32 gene mutation, but that was not proven categorically.

19. Individuals with mutations in one or both of the CCR5 alleles exhibit resistance to the Human Immunodeficiency Virus type 1 (HIV-1), the most common strain of the virus, as HIV-1 uses CCR5 as a co-receptor of infection. Jon Funder Hansen, "CCR5, Delta-32 Mutation and the HIV Infection Pathway," *Microbe Wiki*, Last revised May 9, 2015, https://microbewiki.kenyon.edu/index.php/CCR5,_Delta-32_Mutation_and_the_HIV_Infection_Pathway

Chapter Five

1. Brener, *Walking through Walls and other Impossibilities*.
2. By projecting screen memories and by neural manipulation during the *mindscan* procedure, the abductors are able to re-constitute abductees' memories and make them think and believe all sorts of things which are not true, including—for the exceptionally deluded and narcissistic personality—that they themselves bear a "message for humanity" from the aliens. These beliefs are inculcated purely to assist the abductors in their work and gain compliance from abductees.
3. *The Day the Earth Stood Still* (1951), https://archive.org/details/The.Day.The.Earth.Stood.Still1951
4. Patrick Gross, "UFOs at close sight: who is who, Donald E. Keyhoe," https://ufologie.patrickgross.org/bio/keyhoe.htm
5. Lorenzen and Lorenzen, *Abduction*.
6. "Jacques Vallee: Scientist, Author, High-Tech Investor," https://www.jacquesvallee.net/
7. Miller, "Medical Procedural Differences".
8. Ibid.
9. Hopkins and Rainey, *Sight Unseen*, 362–68. A kind of EES technique being utilized on two different male abductees is described.
10. Smith, *Chosen*, 80–82. An incident is vividly recalled by a male abductee being shown "his" child gestating in a tank of nutrient fluid. The abductors instructed him (he is called 'John' in the text) to "love" the child, telling him it was important for him to do this and transmit this love to the child.
11. Swann, *Penetration*.
12. Jacobs, *Secret Life* and *The Threat*.
13. "Official Secrets Act 1989," *Wikipedia*, https://en.wikipedia.org/wiki/Official_Secrets_Act_1989

14. In the UK, the original Official Secrets Act of 1889 was amended in 1911, and then several times subsequently. The latest version, of 1989, includes several new provisions and legally defined penalties for violation of trust, the most onerous of which is two years' imprisonment following conviction by jury trial.

15. Pritchard et al., eds., *Alien Discussions*.

16. "Stanton Friedman – Physicist, Lecturer, UFO Researcher," https://www.stantonfriedman.com/index.php

17. Robert Hastings, "Has the British Ministry of Defence released all of its UFO files?," June 22, 2013, https://ufohastings.com/articles/has-the-british-ministry-of-defence-released-all-of-its-ufo-files

18. Hastings and Jacobs, "Repercussions," in *Confession*.

19. "Ex-Air Force Personnel: UFOs Deactivated Nukes," CBS News, September 28, 2010, https://www.cbsnews.com/news/ex-air-force-personnel-ufos-deactivated-nukes; CNN Interview of Robert Hastings following the National Press Club news conference, September 29, 2010, https://www.youtube.com/watch?v=7zXG8jPQWf0

20. Hastings, *UFOs and Nukes*.

Chapter Six

1. Hopkins, *Art, Life and UFOs*.
2. Ibid.
3. Ibid.
4. Ibid.
5. Ibid.
6. Ibid.
7. *Sun Black V*, painted in 1969, is the fifth work of the series begun following Hopkins' 1964 daylight UFO sighting. https://www.pinterest.co.uk/pin/budd-hopkins--599260294165087368/
8. Bloecher, Hopkins, and Stoehrer, "The Stonehenge Incidents – January 1975." A link to the report is provided in the bibliography.
9. Hopkins, *Art, Life and UFOs*.
10. Ibid., 217–231.
11. See Chapter Seven.

12. Hopkins, *Witnessed*, 171. Part Three of this of this remarkable study explores this phenomenon.
13. Collings and Jamerson, *Connections*. Fundamentally, the whole book explores the complexity of this deep but little-researched phenomenon.
14. Bryan, *Close Encounters of the Fourth Kind*. The last section of this long and excellent journalistic report on the abduction phenomenon is dedicated to interviewing and exploring this case. The co-authors of Connections are named 'Carol' and 'Alice' in Bryan's account.
15. The International Center for Abduction Research (ICAR) site is no longer active, as David Jacobs has retired from the field of abduction research. https://www.ufoabduction.com/
16. "Timothy Good – UFO Authority," http://www.timothygood.co.uk/
17. "Jerome Clark," n.d., *Wikipedia*, https://en.wikipedia.org/wiki/Jerome_Clark
18. Hopkins, *Art, Life and UFOs*.
19. Budd Hopkins, unpublished journal, 2011.
20. Ibid.

Chapter Seven

1. Jacobs, *The UFO Controversy in America*.
2. Jacobs, "Evolution of an Abduction Researcher," in *Walking Among Us*.
3. Jacobs, *The Threat*.
4. Ibid.
5. Ibid.
6. Ibid.
7. Ibid.
8. Ibid.
9. Ibid.
10. Ibid.
11. "Temple University," https://www.temple.edu/
12. The American Philosophical Society was founded by Benjamin Franklin, holds one of the largest Darwin collections outside Cambridge, and held the likes of

Pasteur, Edison, Darwin, Agassiz, Pauling, Audubon and Mead as its members: pretty illustrious company. https://www.amphilsoc.org/

13. "Charles Library | Temple University," https://www.temple.edu/about/libraries/charles

14. "David M. Jacobs | UFO Abduction Research," https://www.davidmichaeljacobs.com/

Chapter Eight

1. "A growing number of mental-health professionals have conducted their own investigations of the abduction phenomenon. In the early 1980s psychological testing of a small group of abductees in New York indicated that they suffered from post-traumatic stress disorder (PTSD). Dr Elizabeth Slater, a psychologist with a private practice in New York City, remarked that these findings are "not inconsistent with the possibility that reported UFO abductions have, in fact, occurred." Other studies since then have come to similar conclusions, and the scientific investigation of the abduction phenomenon continues." HowStuffWorks, "Alien Abductions," February 18, 2008, https://science.howstuffworks.com/space/aliens-ufos/alien-abduction.htm

2. David M. Jacobs, "The Basics, For Debunkers," accessed June 18, 2022, https://www.davidmichaeljacobs.com/2021/08/05/thinking-clearly/

3. Ibid.

4. Pritchard et al., eds., *Alien Discussions*.

5. "Historical Estimates of World Population," US Census Bureau, Last revised December 16, 2021, https://www.census.gov/data/tables/time-series/demo/international-programs/historical-est-worldpop.html

6. "Population of United States of America 1991," Population Pyramid, accessed June 18, 2022, https://www.populationpyramid.net/united-states-of-america/1991/

7. Jacobs, *The Threat*, 122–25.

8. Hopkins, "Other Women, Other Men," in *Intruders*.

9. Lindemann, ed., *UFOs and the Alien Presence*.

Chapter Nine

1. "George Gamow," *Wikipedia*, accessed June 18, 2022, https://en.wikipedia.org/wiki/George_Gamow

2. Dr Christopher Palma, "The Age of the Universe," Penn State College of Mineral Sciences, accessed June 18, 2022, https://www.e-education.psu.edu/astro801/content/l10_p5.html1

3. Leonor Sierra, "Are we alone in the universe? Revisiting the Drake equation," NASA Exoplanet Exploration, May 19, 2016, https://exoplanets.nasa.gov/news/1350/are-we-alone-in-the-universe-revisiting-the-drake-equation/

4. Maggie Masetti, "How many Stars in the Milky Way?," NASA Blueshift, July 22, 2015, https://asd.gsfc.nasa.gov/blueshift/index.php/2015/07/22/how-many-stars-in-the-milky-way/y

5. Cooper et al., "Glowing Auras and 'Black Money'"

6. Masetti, "How many Stars in the Milky Way?"

7. Dr Christopher Palma, "The Drake Equation," Penn State College of Mineral Sciences, accessed June 18, 2022, https://www.e-education.psu.edu/astro801/content/l12_p5.html

8. "South Park, Season 1, Episode 1: Cartman Gets An Anal Probe," *Watch Cartoons Online*, https://www.wcostream.com/playlist-cat/47948/south-park-season-1-episode-1-cartman-gets-an-anal-probe-2

9. Some abductees tell me they seem mildly repelled by strong perfume. If true, there must be some olfactory function present, maybe through the skin.

10. Such beings also make an appearance in Hopkins and Rainey, *Sight Unseen*.

11. Yurdozu, *Confessions of a Turkish Ufologist*. [emphasis mine]

12. Ibid.

13. "Hormones During Pregnancy," Johns Hopkins Medicine. November 19, 2019, https://www.hopkinsmedicine.org/health/conditions-and-diseases/staying-healthy-during-pregnancy/hormones-during-pregnancy

14. Brent Swancer, "The Hard Road to Evidence of a Human Soul," *Mysterious Universe*, March 10, 2018, https://mysteriousuniverse.org/2018/03/the-hard-road-to-evidence-of-a-human-soul; Robert Lanza, MD, "Does the Soul Exist? Evidence Says 'Yes'," *Psychology Today*, December 21, 2011, https://www.psychologytoday.com/us/blog/biocentrism/201112/does-the-soul-exist-evidence-says-yes

15. I had a conversation with David Jacobs some years ago during which I asked him if he knew or had any information as to whether or not female hubrids menstruated regularly or were fertile and able to conceive like a normal human female. He had no reports of this from any abductee reporting regular contact/

relations with hubrids, so admitted he did not know the answer to this question. All that is known for certain is that hubrids share no mutual attraction and *never mate with each other*, as this activity would not serve the objectives of the program.

16. For example, the theoretical ideas about faster-than-light and interdimensional travel explored in *Hyperspace* and *Parallel Worlds* by Michio Kaku.

17. Zaini Majeed, "Russia's TASS Confirmed Voronezh UFO Incident in 1989," *Republic World*, last revised October 9, 2020, https://www.republicworld.com/technology-news/science/russias-tass-confirmed-voronezh-ufo-incident-on-this-day-in-1989.html

18. "Freedom of Information Act 2000," https://www.legislation.gov.uk/ukpga/2000/36/contents

19. Timothy Good also acquired a reputation, later in life, for embracing some claims seen as less credible by the UFO field, such as those of 'contactee' the late George Adamski and the now deceased Italian writer Stefano Breccia's prolix and elaborate tales about the Ummo and Amicitzia visitations in his book, *Mass Contacts*. Tim, born in 1942, is now retired from the field and officially non-contactable through any channel excepting his literary agent, Andrew Lownie: https://www.andrewlownie.co.uk

20. There are precedents even in human civilizations of multigenerational projects. Though not comparable to the abduction/hybridization program in any other respect, building each of the Gothic cathedrals in Europe through the 13th and 14th centuries spanned the lifetimes of several generations of masons and other workers, most of whom saw neither the initiation nor the conclusion of the project during their natural lifetimes though they spent their own working lives fully engaged with the construction. Building the Great Wall along China's northern border took even longer.

Epilogue

1. "UFOs have taken U.S. nuclear capabilities 'offline,' says former AATIP director," *Washington Post Live*, https://www.youtube.com/watch ?v=tioJj_lqtLU

2. John Greenwald, "Pentagon Now Admits AATIP Utilized UAP/UFO Reports," *Black Vault*. May 22, 2021, https://www.theblackvault.com/documentarchive/pentagon-now-admits-aatip-utilized-uap-ufo-reports/; "George Knapp & Luis Elizondo – UFOs: What's Next? … Coast To Coast AM 2022," YouTube: Coast To Coast AM Official, https://www.youtube.com/watch?v=vHPJ1jgKt7c

Appendix A

1. "An increased number of fibrocytes and histocytes is accompanied by coarse bundles of collagen arranged randomly". A useful reference for the reader who has a basic working understanding of cytology might be: Adriana Blakaj and Richard Bucala, "Fibrocytes in health and disease," *Fibrogenesis Tissue Repair.* 2012; 5(Suppl 1): s6. Published online June 6, 2012, https://www.ncbi.nlm.nih.gov/pmc/articles/PMC3368795/

Appendix C

1. Microtrace LLC is a microscopy and microchemistry forensic consulting laboratory based in Illinois. www.microtrace.com

Acknowledgements

I wish to acknowledge the generosity and assistance of the many people who, directly or indirectly, contributed to the body of knowledge and understanding which enabled this book to be born into the world.

Without the kindness and active assistance of Nick Pope, the door to understanding this phenomenon might have never been opened and the journey of discovery started in 2007 never begun. Nick's generosity, valuable support and continued encouragement during the early years of my engagement with this subject was always exceptional. His friendship and support continues to this day.

The invaluable encouragement of Robert Hastings inspired the confidence necessary to undertake writing the manuscript. In addition to his sage counsel regarding the subject matter of this book, Robert provided invaluable advice about editing, publishing and distribution, and for that I will be forever thankful. Robert's close friendship over the previous 12 years is dear to me. Our long discussions about every aspect of this phenomenon have been a joy and a source of endless fascination.

As a fan of Jared Tarbell's distinctive, computer-generated fractal designs, I approached him to create some unique cover-art for *Out of Time*. Jared offered many innovative ideas and one of his Fractured Disc images was selected to grace the book's cover. My gratitude to Jared for offering to this project a most striking, unique cover design.

Frances Watts generously assisted in researching the family line. Her experience and knowledge of genealogy helped greatly – if unwittingly - in focusing the narrative on the ancestral line, which has proved to be so central to the thesis of *Out of Time*.

My editor Brian Ally was tireless in the thoroughness of his efforts to fashion my manuscript into shape and transform it into a sharper and more focussed narrative. He also guided me through the esoteric world of typesetting and publishing and was above all the indispensable key to grappling with and understanding essential elements of the publishing world.

Peter Robbins was supportive in innumerable ways. He connected me with Budd Hopkins, so propelling me into the rarefied environment where I was able to mix freely with the most incisive and ground-breaking minds engaged with the UFO and abduction research community. He was always steadfast in his encouragement and gave invaluable feedback to an early iteration of the manuscript.

Peter's endorsement of my work persuaded Ralph Blumenthal to give time to my fledgling manuscript. Ralph's exceptional knowledge of the abduction mystery, due to his long research into the life and work of Dr. John Mack, was unquestionably a significant help to me in including my own examination of Dr. Mack's life and work in *Out of Time*. Furthermore, Ralph's encouragement after reading an early iteration of the manuscript motivated my decision to publish many of the more controversial ideas expounded in the book.

My indebtedness to David Jacobs will be lifelong. His selfless devotion to abductees, and the almost limitless time he offered freely, enabled many hidden memories to emerge, both while he worked with me and afterwards. Our extensive dialogues regarding the abduction phenomenon were above all the most valuable asset which I held when I first decided to write *Out of Time*, following his retirement from this field of study in 2018. He also kindly offered corrective suggestions to some details of his bio detailed in Chapter Seven.

I will always owe a debt of gratitude to the late Budd Hopkins, who during his final years with us shared with me some of his unpublished research, including a great deal about the lives of many abductees which he had come to understand but which remains unpublished and otherwise unrecorded. His encouragement of my fledgling literary efforts and compliments to my writing style served as a great confidence booster towards the eventual writing of this book.

Dr. Don Donderi, having read the manuscript of *Out of Time*, graciously offered to write a *Foreword* to the book, so lending his five-star professional and academic credentials to what might have otherwise remained the largely unknown efforts of an ill-qualified and hitherto unpublished writer. For this valuable contribution I express my gratitude.

Finally but by no means least, the continued encouragement of my wife Janis for me to record my lifelong history of anomalous experiences and write this book has been of inestimable value. Her patience, honesty and constructive criticism during the many months of this project, and her technical expertise with IT which far exceeds my own, have all contributed in great measure to the finished work. It exists in no small part due to her.

About the Author

Steve Aspin has a professional background in marketing, sales, and business management and worked for thirty-five years in the fields of surgery and medical diagnostics. In 1999 he founded what became an internationally successful surgical innovations company based near London. He has designed, patented, manufactured, and successfully exported thousands of innovative products for both surgery and medical diagnostics to the global marketplace.

Out of Time is his first published book on the subject of UFO and abduction research, following a lifetime of coerced entanglement with this phenomenon. He is now retired, lives with his wife in eastern England, and may be found most early mornings walking his dog across the Lincolnshire Wolds.

Printed in the USA
CPSIA information can be obtained
at www.ICGtesting.com
LVHW020903111123
763663LV00044B/1805